ENERGY SCIENCE, ENGINEERING AND TECHNOLOGY SERIES

INTRODUCTION TO POWER GENERATION TECHNOLOGIES

ENERGY SCIENCE, ENGINEERING AND TECHNOLOGY SERIES

Oil Shale Developments
Ike S. Bussell (Editor)
2009. ISBN: 978-1-60741-475-9

Power Systems Applications of Graph Theory
Jizhong Zhu
2009. ISBN: 978-1-60741-364-6

Bioethanol: Production, Benefits and Economics
Jason B. Erbaum (Editor)
2009. ISBN: 978-1-60741-697-5

Introduction to Power Generation Technologies
Andreas Poullikkas
2009. ISBN: 978-1-60876-472-3

ENERGY SCIENCE, ENGINEERING AND TECHNOLOGY SERIES

INTRODUCTION TO POWER GENERATION TECHNOLOGIES

ANDREAS POULLIKKAS

Nova Science Publishers, Inc.
New York

For permission to use material from this book please contact us:
Telephone 631-231-7269; Fax 631-231-8175
Web Site: http://www.novapublishers.com

LIBRARY OF CONGRESS CATALOGING-IN-PUBLICATION DATA

Poullikkas, Andreas.
Introduction to power generation technologies / author, Andreas Poullikkas.
p. cm.
Includes bibliographical references and index.
ISBN 978-1-60876-472-3 (hardcover)
1. Electric power production. I. Title.
TK1001.P596 2009
621.31'21--dc22

2009032474

Published by Nova Science Publishers, Inc. ✢ New York

CONTENTS

PREFACE

Introduction to Power Generation Technologies provides an insight to the wide range of electricity generating technologies available today or under development. Also, provides an overview of the energy storage technologies which are expected to play an important role in the future sustainable energy systems. The technologies are presented in an easily digestible form. The book gives a clear, unbiased review and comparison of the different types of power generation technologies available.

The book is divided into seven chapters as follows:

Chapter 1: *Introduction*: including the basic laws of thermodynamics, the different fossil fuels and their combustion, the related primary emissions and greenhouse gas emissions, as well as, a background on alternative energy sources and the future sustainable energy systems,

Chapter 2: *Power plants*: analysis of the most important conventional technologies for power generation that are currently used or under development, such as, the Rankine cycle, the simple cycle gas turbine, the combined cycle technologies, nuclear power, etc.,

Chapter 3: *Carbon capture and storage technologies*: an overview of the most common power generation technologies which can incorporate carbon capture and storage systems, such as pulverised coal technologies, natural gas combined cycle technologies, etc.,

Chapter 4: *Direct solar RES technologies*: analysis of solar thermal power generation technologies and photovoltaic systems,

Chapter 5: *Indirect solar RES technologies*: an overview of indirect solar RES technologies, such as, wind, biomass, hydropower, etc., including advantages, disadvantages and future technological developments,

Chapter 6: *Distributed generation*: including definition and overview of fuel cells technology,

Chapter 7: *Storage technologies*: an overview of various energy storage technologies currently in use or under development, such as, flywheels, batteries, hydrogen storage devices, etc.

Introduction to Power Generation Technologies is not by any means exhaustive, nor is it intended to be. In the more than two decades I've worked with power generation technologies, the field has grown so vast that it's no longer possible to confine the technologies within the covers of one book, even after limiting it to the most important systems.

The book is partly based on lecture notes provided in two different courses for a number of years and is intended as an introductory textbook for courses in the field of engineering, environmental pollution and public health. Also, *Introduction to Power Generation Technologies* can serve as a reference text for power generation planners, electric utility managers, energy regulators, electricity transmission system operators, consultants, policy makers and economists.

This book is dedicated to my wife Rania and daughter Agnes whose love and support sustained me throughout the preparation of this book.

Dr. Andreas Poullikkas
Nicosia, Cyprus
2009

ABOUT THE AUTHOR

Dr. Andreas Poullikkas holds a B.Eng. degree in mechanical engineering, an M.Phil. degree in nuclear safety and turbomachinery, and a Ph.D. degree in numerical analysis from Loughborough University, UK. His present employment is with the Electricity Authority of Cyprus where he holds the post of Assistant Manager of Research and Development.

Dr. Poullikkas is, also, a Visiting Fellow at the University of Cyprus and at the Cyprus International Institute for the Environment and Public Health in association Harvard School of Public Health, USA. In his professional career he has worked for academic institutions, before joining the Electricity Authority of Cyprus. He has over 20 years experience on research and development projects related to the numerical solution of partial differential equations, the mathematical analysis of fluid flows, the hydraulic design of turbomachines, the nuclear power safety, the analysis of power generation technologies and the power economics.

Dr. Poullikkas is the author of various peer reviewed publications in scientific journals, book chapters and conference proceedings. He is, also, a referee for various international journals, serves as a reviewer for the evaluation of research proposals related to the field of energy and a coordinator of various funded research projects. He is a member of various national and European committees related to energy policy issues. He is the developer of various algorithms and software for the technical, economic and environmental analysis of power generation technologies, desalination technologies and renewable energy systems.

Chapter 1

1. INTRODUCTION

The availability of energy and people's ability to harness that energy in useful ways has transformed our society. A few hundred years ago, the greatest fraction of the population struggled to subsist by producing food for local consumption. Now, in many countries a small fraction of the total workforce produces abundant food for the entire population and much of the population is thus freed for other pursuits. We are able to travel great distances in short times by using a choice of conveyances (including trips to earth orbit as well as to our nearest natural satellite). We can communicate instantaneously with persons anywhere on earth and we control large amounts of energy at our personal whim in the form of automobiles, electric tools and appliances, and comfort conditioning in our dwellings.

These changes resulted from a combination of inventiveness and ingenuity, coupled with a painstaking construction of theory by some of the great scientist and engineers throughout the years. As a result of the science and application of thermodynamics, our ability to obtain energy, transform it and apply it to society's needs has brought about the change from agrarian to modern society.

Because of its generality, thermodynamics is the underlying science that forms the framework for the study of most other engineering subjects. The most obvious are heat transfer, the study of how energy is transferred from a material or location at a certain temperature to another material or location at a different temperature, and fluid mechanics, which deals with the motion of fluids under externally applied forces and the transformation of energy between mechanical and thermal forms during this motion.

Another way of seeing pervasiveness of thermodynamics in studies of interest to engineers is to examine the many and diverse areas of application. These include power plants (fossil fuel, nuclear fission, nuclear fusion, solar, thermal, geothermal, etc.), engines (steam, gasoline, diesel, stationary and propulsion gas turbines, rockets, etc.), air conditioning and refrigeration systems, furnaces, heaters, chemical process equipment, the design of electronic equipment (for example, to avoid overheating and failure of individual components, circuit boards, etc.), design of mechanical equipment (for example in lubrication of bearings to predict the overheating and subsequent failure due to excessive applied loads and in break design to predict lining wear rates due to frictional heating and erosion) and in manufacturing processes (for example, the wear of tool bits is often due to frictional heating of the cutting edge). Indeed, it is fairly easy to make a case that thermodynamics in its broadest sense is the underlying science in most fields of engineering. Even the fields of pure mechanics use energy conservation relations which are subsets of more general thermodynamic principles.

1.1. Definition of Thermodynamics

Thermodynamics is defined as the study of energy, its forms and transformations, and the interaction of energy with matter. The most important application of thermodynamics concerns the conversion of one form of energy into another, especially the conversion of heat into other forms of energy. These conversions are governed by the two fundamental laws of thermodynamics. The first of these is essentially a general statement of the law of conservation of energy and the second is a statement about the maximum efficiency attainable in the conversion of heat into work.

The study of thermodynamics was inaugurated by nineteenth century engineers who wanted to know what ultimate limitations the laws of physics impose on the operation of steam engines and other machines that generate mechanical energy. They soon recognized that perpetual motion machines are impossible. A perpetual motion machine of the first kind is a (hypothetical) device that supplies an endless output of work without any input of fuel or any other input of energy. As we will see later, the First Law of Thermodynamics, or the law of conservation of energy, directly tells us of the failure of this machine since after one revolution of the wheel, the masses all return to their initial positions, their potential energy returns to its initial value, and they will not have delivered any net energy to the motion of the wheel.

A perpetual motion machine of the second kind is a device that extracts thermal energy from air or from the water of the oceans and converts it into mechanical energy. Such a device is not forbidden by conservation laws. The oceans are an enormous reservoir of thermal energy; if we could extract this thermal energy, a temperature drop of just 1°C of the oceans would supply the energy needs of the United States for the next 50 years. But, as we will see, the Second Law of Thermodynamics tells us that conversion of heat into work requires not only a heat source, but also a heat sink. Heat flows out of a warm body only if there is a colder body that can absorb it. If we want heat to flow from the ocean into our machine, we must provide a low-temperature heat sink toward which the heat will tend to flow spontaneously. Without a low-temperature sink, the extraction of heat from the oceans is impossible. We cannot build a perpetual motion engine of the second kind.

1.2. Energy Balance Approach

The principle "energy can be neither created nor destroyed" is one of the conservation relations which, when carefully expanded and explored, forms the basis for a good deal for the study of thermodynamics. The conservation of energy principle can be made true in any situation by simply changing or redefining what we mean by energy, so that it is indeed conserved in all situations. The basic conservation principle has two important suppositions. The first is that energy is something that is "contained". A certain defined system has "energy". The second supposition is that there is a well specified system where the energy is contained. To apply the energy conversation principle the user must define the space or material of interest that "has" the energy.

Without worrying at this point about how energy is classified we can apply the conversation of energy principle to a power plant used for the production of electricity. For example a power plant uses 1 unit of fuel energy to produce 0.4 energy units of electricity. We need to determine the net energy transferred to the environment during the conversion of fuel to electricity.

For this type of problem, the energy inside the power plant boundary remains constant according to the conservation of energy principle (since it cannot created or destroyed). We could write in this case that:

$$\begin{pmatrix} Energy \\ transferred \\ in \end{pmatrix} + \begin{pmatrix} Energy \\ transferred \\ out \end{pmatrix} = 0, \qquad (1)$$

where *in* and *out* refer to the direction in which the energy is crossing the plant boundary. Now we can expand these terms to include each of the energy transfer processes, or

$$\begin{pmatrix} Energy \\ transferred \\ in \end{pmatrix}_{fuel} + \begin{pmatrix} Energy \\ transferred \\ out \end{pmatrix}_{electricity} + \begin{pmatrix} Energy \\ transferred \\ in \end{pmatrix}_{\substack{cooling \\ water}} + \begin{pmatrix} Energy \\ transferred \\ out \end{pmatrix}_{\substack{cooling \\ water}} + \begin{pmatrix} Energy \\ transferred \\ out \end{pmatrix}_{\substack{stack \\ gases}} = 0 , \quad (2)$$

or

$$1 \text{ unit} - 0.4 \text{ unit} + \begin{pmatrix} Energy \\ transferred \\ in \end{pmatrix}_{\substack{cooling \\ water}} + \begin{pmatrix} Energy \\ transferred \\ out \end{pmatrix}_{\substack{cooling \\ water}} + \begin{pmatrix} Energy \\ transferred \\ out \end{pmatrix}_{\substack{stack \\ gases}} = 0 \quad (3)$$

or finally,

$$\begin{pmatrix} Energy \\ transferred \\ out \end{pmatrix}_{environment} = \begin{pmatrix} Energy \\ transferred \\ in \end{pmatrix}_{\substack{cooling \\ water}} + \begin{pmatrix} Energy \\ transferred \\ out \end{pmatrix}_{\substack{cooling \\ water}} + \begin{pmatrix} Energy \\ transferred \\ out \end{pmatrix}_{\substack{stack \\ gases}} = -0.6 \text{ unit} \quad (4)$$

This very simple analysis illustrates a number of points about the energy conservation principle. First, we must carefully defined the location to which the principle will be applied, in this case the power plant. Second we must define a convention for how we assign the sign of the energy transfers. Here we simply chose energy transfers into the plant as carrying a positive sign, so that energy transferred out is negative. Finally we have tacitly kept the units (dimensions) of each quantity in the energy balance consistence.

In more practical applications the energy transfers in the various terms of the energy balance have different forms. For example, in the problem above, the energy transferred it might be chemical energy in a fossil fuel such as coal, oil or natural gas. It might be the binding energy of the nuclei of atoms for a nuclear plant or the energy transfer from the sun in a solar power plant. The electric energy is in the form of an electric current carried by the transmission lines leaving the plant. The energy transfer to the cooling water is usually in the form of thermal energy added to cooling water or the atmosphere, which then leaves the plant. Finally, the energy from the stack is transported in the hot flowing gas from the plant to the atmosphere. Thus, we need to classify the energy transfer across the plant boundary so that we can properly set up our energy accounting system.

One additional point is that we assumed (without stating the assumption) that the energy entering the plant was exactly balanced at each instant by the energy leaving. However, this is not always the case. Consider a new system boundary for a coal-fired power plant that includes a coal stockpile, where coal is often delivered to the plant for later use at periods of high electrical demands. In such a case, our energy conservation equation must be extended to account for an energy storage term. Alternatively, the boundary could be redrawn to include only the powerhouse itself, excluding the coal pile, so that the original conservation equation could still be used.

1.3. Heat and Work

Heat, Q, is a form of energy which is transferred from one body to another body at a lower temperature by virtue of the temperature difference between the bodies.

Referring to Figure 1, provided the temperature, T, of body A is greater than the temperature of body B, i.e., $T_A > T_B$, heat is flowing from body A to body B until thermal equilibrium is reached, that is $T_A = T_B$. *Work*, W, is defined as the product of a force, F, and distance, d, moved in the direction of the force, as indicted in Figure 2. In mathematical form,

$$W_{AB} = F \times d \tag{5}$$

Both heat and work can never be contained in a body or possessed by a body.

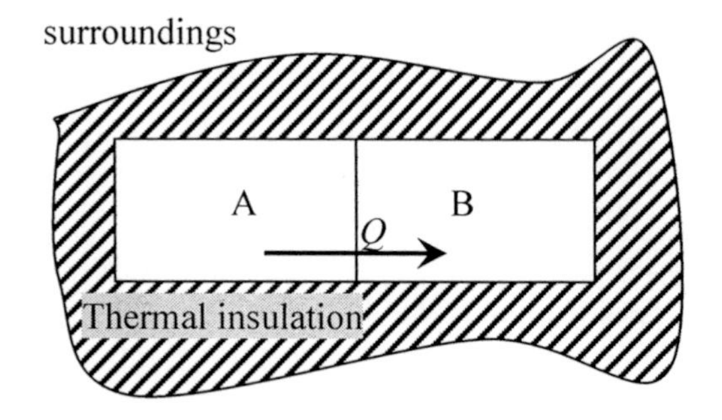

Figure 1. Heat transfer from body A to body B.

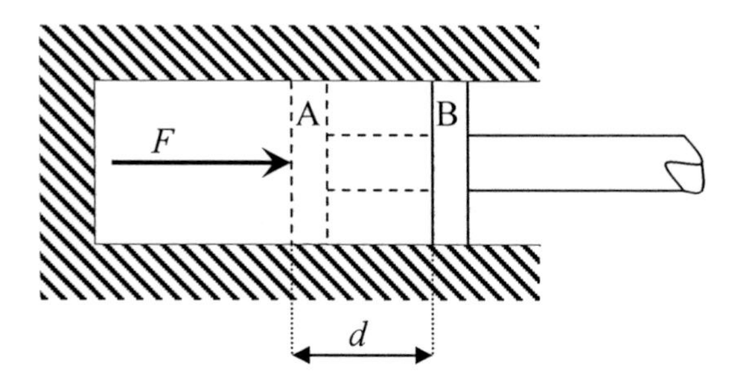

Figure 2. Work from point A to point B.

A *system* may be defined as a collection of matter within prescribed and identifiable boundaries. A closed system is presented in Figure 3, in which there is no mass transfer across the boundary.

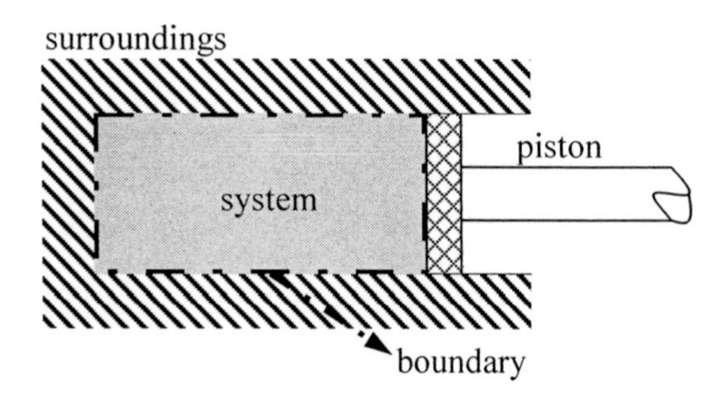

Figure 3. Closed system.

An open system is one where material can flow across the boundaries of the system, as well as heat and work. In the process shown in Figure 4, gas is entering the system and being heated in the heat exchanger where its temperature pressure and volume increase. The gas then passes through an expansion turbine, where it cools down and the pressure decreases. During the expansion process work is extracted from the gas.

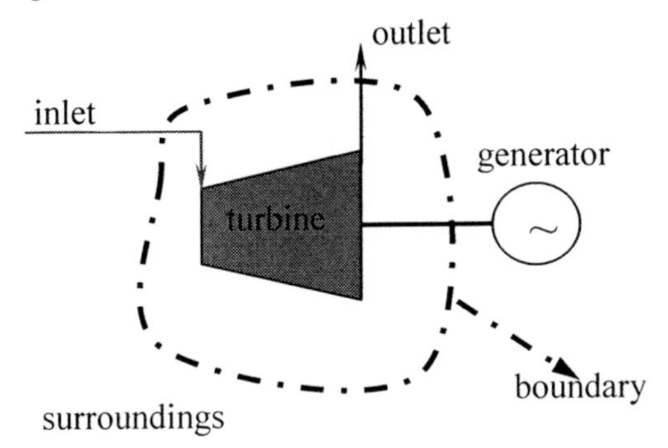

Figure 4. Open system (mass transfer).

1.4. First Law of Thermodynamics

When a system undergoes a thermodynamic cycle then the net heat supplied to the system from its surroundings is equal to the net work done by the system on its surroundings. We can express this change of energy as:

$$\Sigma\, dQ = \Sigma\, dW \qquad (6)$$

Note that the first law simply states that energy is conserved, that is the change of the internal energy equals the input of heat and work. By this the unknown amount of heat or work required for a process, if the amount of heat and work for a different process that takes the system from the same initial state to the same final state are known can be calculated.

1.4.1. The Non-Flow Equation

Consider a process or a series of processes between state 1 and state 2 provided there is no flow of fluid into or out of the system as indicated in Figure

3. The intrinsic energy of the system is finally greater than the initial intrinsic energy, that is, the gain in intrinsic energy equals the net heat supplied minus the net work output. The non-flow equation is given algebraically by:

$$Q = (U_2 - U_1) + W\ , \tag{7}$$

where Q is the heat in kJ, U is the internal energy in kJ and W is the work in kJ.

1.4.2. The Steady Flow Energy Equation

When 1 kg of a fluid with specific internal energy, u, is moving with a velocity C and is at a height Z above datum level as indicated in Figure 5, then it possesses a total energy of $u + (C^2/2) + Zg$ where $(C^2/2)$ is the kinetic energy of 1kg of the fluid and Zg is the potential energy of the fluid.

Since there is a steady flow of fluid into and out of the system and there are steady flows of heat and work, then the energy entering must exactly equal the energy leaving:

$$u_1 + \frac{C_1^2}{2} + Z_1 g + P_1 \upsilon_1 + q = u_2 + \frac{C_2^2}{2} + Z_2 g + P_2 \upsilon_2 + w\ , \tag{8}$$

where u is the specific internal energy in kJ/kg, C is the velocity in m/s, Z is the level height in m, g is the acceleration due to gravity in m/s^2, P is the pressure in N/m^2, υ is the specific volume in m^3/kg, q is the specific heat in kJ/kg and w is the specific work in kJ/kg.

Also, the potential energy is negligible and the enthalpy, h, is defined as $h = u + P\upsilon$, therefore,

$$h_1 + \frac{C_1^2}{2} + q = h_2 + \frac{C_2^2}{2} + w\ . \tag{9}$$

The above is known as the steady flow energy equation.

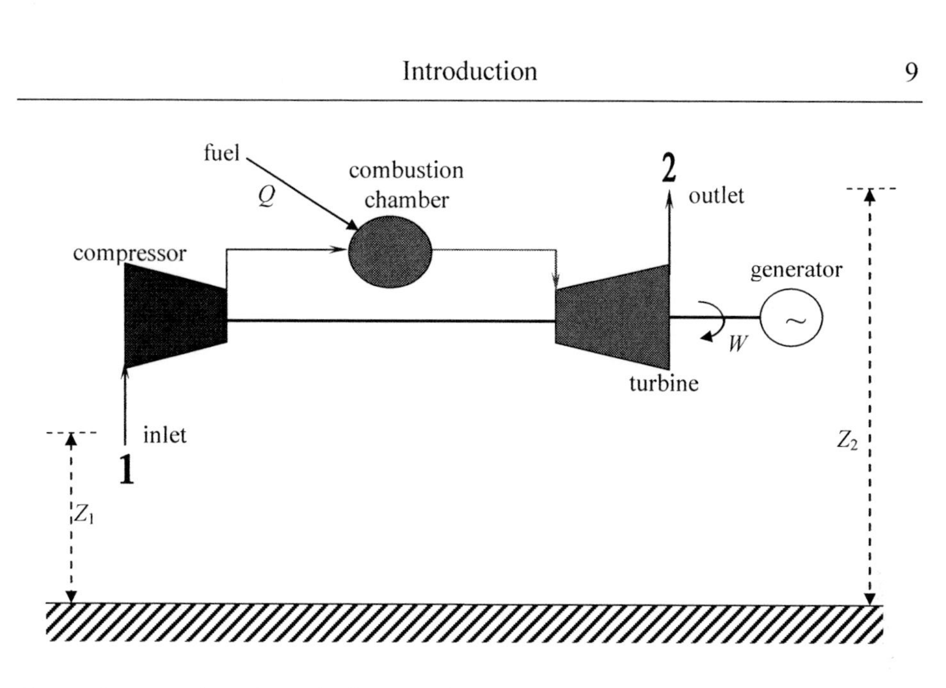

Figure 5. A complete schematic diagram of an open system.

1.5. Second Law of Thermodynamics

The gross heat supplied must be grater than the net work done; some heat must be rejected by the system. We can express this as:

$$Q_1 > W\ , \tag{10}$$

where Q_1 is the heat supplied from the source and W is the net work done.

The second law of thermodynamics is a statement about the maximum efficiency attainable in the conversion of heat into work, that is, the conversion of heat into work requires not only a heat source, but also a heat sink.

Referring to Figure 6, in which Q_1 is the heat supplied from the source, Q_2 is the heat rejected and W is the net work done, by applying the first law of thermodynamics we obtain:

$$\sum dQ = \sum dW \text{ or } Q_1 - Q_2 = W\ . \tag{11}$$

The thermal efficiency η is then given as:

$$\eta = \frac{W}{Q_1} = \frac{Q_1 - Q_2}{Q_1} = 1 - \frac{Q_2}{Q_1}. \quad (12)$$

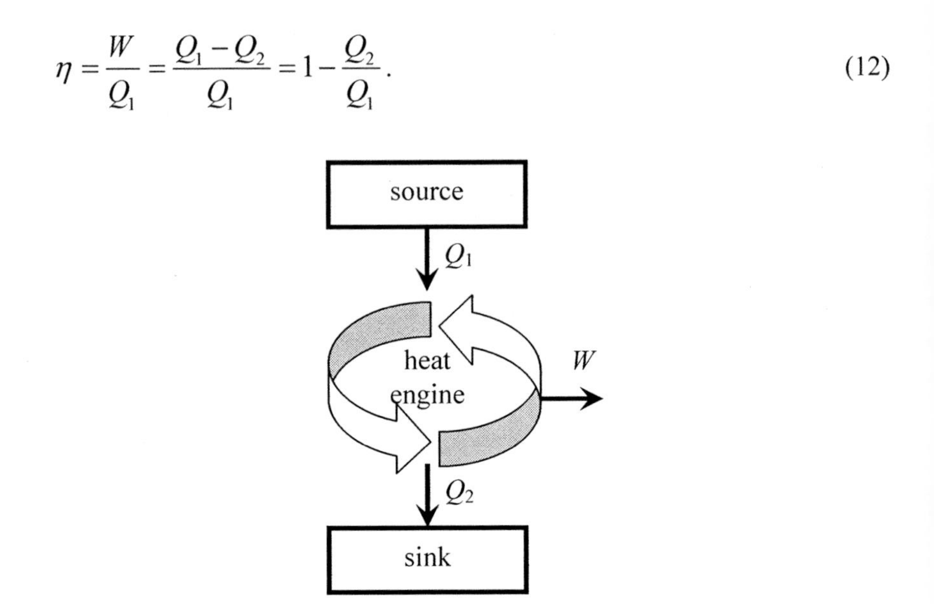

Figure 6. The principle of second law of thermodynamics.

It can be seen that second law implies that the thermal efficiency of heat engine must always be less that 100%.

A heat engine, shown in Figure 7, is a system operating in a complete cycle and developing net work from a supply of heat. In particular liquid enters the pump and is increased in pressure, work input W_2, before entering the boiler at relatively low temperature.

The energy added to the fluid, heat supplied Q_1, from the combustion process of the fuel rises the temperature of the fluid producing vapor. Then, the vapor is expanded through a turbine to produce work, work output W_2. In the end of the expansion process, low temperature vapor leaves the steam turbine. The vapor is then condensed to a liquid as the vapor comes into contact with surfaces in the condenser that are cooled, heat rejection Q_2, usually by a cold water stream. Following condensation the liquid enters a pump. The working fluid is returned to the high pressure necessary for energy addition at the higher boiler temperature and the cycle is repeated.

The net work done is given by:

$$W = W_1 - W_2. \quad (13)$$

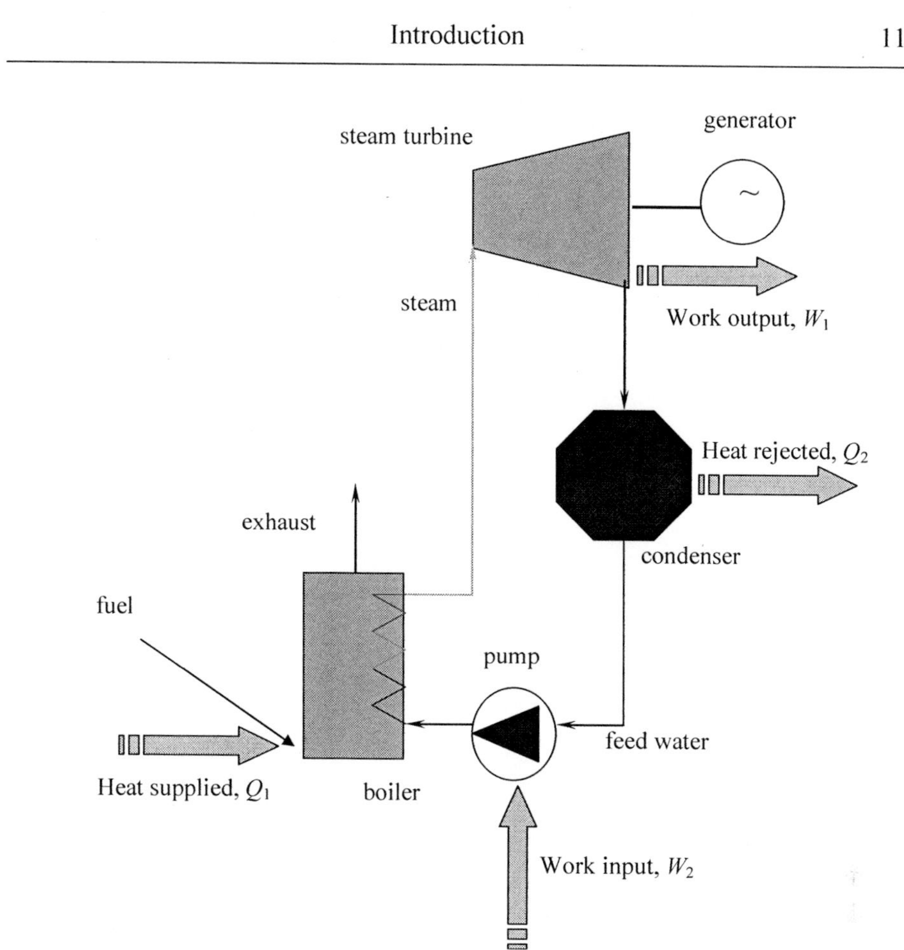

Figure 7. A schematic diagram of a heat engine.

1.6. FUELS

Coal fired power plant capacity has internationally retained its position as the major source of electricity generation, being remarkably stable at around 38% since at least 1975. With reference to Table 1, the International Energy Agency projects that this share will be maintained up to 2010, with strong growth expected in countries such as India and China. The oil-based share of generation has continued the decline that was initiated by the oil shocks of the 1970s. While nuclear-based power production increased significantly in the 1980s, its share seems now to have peaked following the few, but serious, safety shocks that have precluded new capacity in many countries. Expansion of nuclear capacity, a capital-intensive technology, is also not favored by the higher required rates of return in restructured and privatized electricity sectors.

Table 1. World fuel share for electricity generation

Technology	Fuel share (%)				
	1975	1980	1990	2000	2010
Coal	36.7	38.1	38.2	37.8	38.2
Oil	22.1	19.7	11.3	9.6	8.0
Natural gas	2.5	12.0	13.7	14.8	24.3
Nuclear	5.9	8.6	17.0	17.6	12.3
Renewables	23.0	21.6	19.8	20.2	17.2
Total	90.2	100.0	100.0	100.0	100.0

The natural gas share has increased by 2.5% since 1975 to reach around 15% in 2000. Factors favoring an accelerating increase in the natural gas share have included greatly expanded natural gas reserves and pipeline delivery systems around the world. Policy perceptions have, also, changed during the late 1980s and 1990s. Natural gas now appears not to be a scarce premium fuel subject to supply security risks. Rather, it is now viewed as a moderately low-cost, plentiful and generally secure fuel that is, also, cleaner than other fuels in terms of both primary emissions (SO_X, NO_X, dust) and greenhouse gas emissions (e.g., CO_2). Accordingly, the International Energy Agency projects continued strong growth in its share, overtaking nuclear power to take second place to coal-based electricity by 2010.

1.6.1. Heavy Fuel Oil

Heavy fuel oil (HFO) is one of the by-products resulting from the fractional distillation of crude oil. It is noted that the crude oil can be refined after it has been separated from natural gas and water. HFO consists mainly of carbon and hydrogen known as hydrocarbons and of small quantities of nitrogen, oxygen and sulphur. Also, it consists of other inorganic compounds and elements, such as, salts of sodium, lime, magnesium, iron, etc. A typical stoichiometric analysis of HFO, with 1% sulphur content, used for power generation, is given in Table 2 (the gross and net calorific values are also tabulated).

Table 2. Typical stoichiometric HFO analysis used for power generation

Constituent	Content
Carbon	85.0%
Hydrogen	12.0%
Oxygen	1.0%
Sulphur	1.0%
Nitrogen	0.8%
Ash	0.2%
Calorific value:	
Gross	43 MJ/kg
Net	41 MJ/kg

During HFO combustion, many of the inorganic elements break down forming oxides and remaining into the ash, the volume of which is negligible. Sulphur, which is contained in the form of inorganic and organic compounds, is especially undesirable. During combustion the main volume of sulphur oxidizes into sulphur dioxide and is emitted into the atmosphere with the exhaust gases. There is an international market for the HFO which is made up of many suppliers. Transport and handling are easy and cheaply facilitated. Tanker capacity is in oversupply, shipping costs are low and site unloading is both simple and cheap. Bulk fuel oil tanks are used for storage until use.

1.6.2. Gasoil

Gasoil is a product resulting from the fractional distillation of crude oil. It is mainly composed of carbon and hydrogen which form hydrocarbons and a small number of the organic and inorganic compounds. A typical stoichiometric gasoil analysis used for power generation is given in Table 3 (the gross and net calorific values are also tabulated). There is an international market for the gasoil which is made up of many suppliers. Transport and handling are easy and cheaply facilitated. Shipping costs are low and site unloading is both simple and cheap. Bulk fuel oil tanks are used for storage until use.

1.6.3. Coal

Coal is extracted from coal mining and its combustible part consists mainly of carbon, hydrogen, and oxygen in small quantities. Its non combustible part consists of water and inorganic compounds, such as, carbonates, phosphates, sulphates of iron, calcium, magnesium, potassium and sodium and oxides of aluminum and silicon. Typical stoichiometric coal analysis, with 1% sulphur content, used for power generation is given in Table 4 (the gross and net calorific values are also given). Also, for comparison purposes the limits of different types of coal analysis are tabulated. During coal combustion, many of the inorganic elements break down and form oxides and remain into the ash. Also, 5% of the ash consists of sulphur, which is contained in the form of inorganic and organic compounds. During combustion the main volume of sulphur oxidizes into sulphur dioxide and is emitted into the atmosphere with the exhaust gases.

Table 3. Typical stoichiometric gasoil analysis used for power generation

Constituent	Content
Carbon	85.0%
Hydrogen	12.0%
Oxygen	1.0%
Sulphur	0.3%
Nitrogen	0.8%
Ash	0.005%
Other elements	0.9%
Calorific value:	
Gross	45 MJ/kg
Net	42 MJ/kg

Like oil, coal is a multi-source fuel which is traded in an international market. World reserves of coal are enormous and are still increasing as the development of mines continues. Thus, future availability is ensured and pressure to moderate future prices is constant. Coal is delivered by vessels and costly unloading facilities are required, such as coal jetty. After it has been unloaded it is stored in open spaces until it can be used.

Table 4. Typical stoichiometric coal analysis used for power generation

Constituent	Content	Limits
Carbon	80.0%	50 - 95%
Hydrogen	5.0%	2 - 45%
Oxygen	6.0%	2 - 40%
Sulphur	1.0%	0.5 - 7%
Nitrogen	2.0%	0.5 - 3%
Ash	6.0%	2 - 30%
Calorific value:		
Gross		27 MJ/kg
Net		25 MJ/kg

Table 5. Typical volumetric natural gas analysis used for power generation

Constituent	Content
Methane	93.0%
Ethane }	
Propane }	4.0%
Butane }	
Nitrogen	1.0%
Carbon dioxide	0.4%
Oxygen	0.3%
Hydrogen	0.7%
Other elements	0.6%
Calorific value:	
Gross	50 MJ/kg
Net	45 MJ/kg

1.6.4. Natural Gas

Natural gas is formed by gaseous hydrocarbons which are trapped on the top of petroleum reservoirs. It contains mainly methane and other hydrocarbons, such as, ethane, propane, butane and pentane. It contains, also, in small concentrations nitrogen, carbon dioxide, oxygen, hydrogen and hydrogen sulphide. Typical volumetric analysis of natural gas, used for power generation, is given in Table 5 (the gross and net calorific values are also tabulated).

There is not an international market for natural gas in the sense that it can be bought at short notice based on international price trends. Neither it is possible to contract for variable quantities or supplies for short periods. Supply is established by contracts between the producer and the consumer, based on annual consumption, for over 20 years. Natural gas is transported by pipelines which are constructed for this purpose.

1.6.5. Liquefied Natural Gas

In the case where there is no means of transportation by pipeline, then the natural gas can be liquefied and transported by ship as liquefied natural gas (LNG). Propane and butane can be liquefied at relatively low pressure and at normal temperature. Hence these two elements are separated from the natural gas giving the well known liquefied petroleum gas (LPG). Methane and ethane, however, could not be liquefied under pressure and at normal temperature. They can be liquefied, however, under cryogenic conditions at a temperature of -160°C under atmospheric pressure. To achieve such temperature, humidity, carbon dioxide and hydrogen sulphide are removed from the natural gas in order to avoid the formation of ice. Also, hydrogen sulphide being toxic and corrosive must be removed. Typical volumetric analysis of LNG, used for power generation, is given in Table 6 (the gross and net calorific values are also tabulated).

As with the natural gas, there is not an international market for LNG in the sense that it can be bought at short notice based on international price trends. Neither it is possible to contract for variable quantities or supplies for short periods. Supply is established by contracts between the producer and the consumer, based on liquefaction, transportation and consumption, for over 20 years.

LNG is delivered by insulated ships and expensive unloading facilities are required, such as a jetty and a cryonic pipeline. LNG is stored in special insulated tanks at a temperature of -160°C. For the efficient use of LNG it is important to regasify it. For this reason there is a need for the installation of an LNG regasifigation plant, the cost of which is high. Also, for security reasons, LNG storage facilities must be at least 2km away from residential areas and 1km away from industrial areas.

Table 6. Typical volumetric LNG analysis used for power generation

Constituent	Content
Methane	94.0%
Ethane	5.0%
Other hydrocarbons	0.5%
Nitrogen	0.5%
Calorific value:	
Gross	50 MJ/kg
Net	45 MJ/kg

1.7. COMBUSTION

The term combustion refers to a very rapid reaction usually accompany by a flame which occurs between fuel and oxygen. The energy is stored in the bores between atoms comprising molecules of the fuel. Companying the fuel with oxygen the energy stored is reduced and the product s released to the surroundings as a heat transfer.

Most common fuels consist mainly of hydrogen and carbon and the chemical composition of most common hydrocarbons is C_nH_n (e.g., natural methane CH_4). The combustible elements in a fuel are:

(a) the Carbon (C),
(b) the Hydrogen (H) and
(c) the Sulphur (S),

and the incombustible elements are:

(a) the Nitrogen (N) and
(b) the moisture (H_2O).

First step in a combustion process is the formation of the chemical equation. The chemical equation,

$$0.85\ C + 0.13\ H + 0.02\ S + 0.90\ O_2 + 3.39\ N_2 \rightarrow 0.85\ CO_2 + 0.065\ H_20 + 0.02\ SO_2 + 3.39\ N_2\ ,$$

represents the chemical balance occurring during the combustion of 2% sulphur content HFO at stoichiometric conditions, neglecting dissociation. The above equation shows how sulphur dioxide is produced through the burning of fossil fuels. In real life, taking into account dissociation, combustion of the same fuel at stoichiometric conditions will give,

$$0.85\ C + 0.13\ H + 0.02\ S + 0.90\ O_2 + 3.39\ N_2 \rightarrow$$
$$0.84\ CO_2 + 0.01\ CO + 0.065\ H_20 + 0.017\ NO_X + 0.02\ SO_2 + 3.38\ N_2\ ,$$

in which we observe that at certain flame conditions a small amount of carbon is partially oxidized to CO and a small amount of nitrogen is oxidized to give NO_X emissions.

1.8. Pollution and the Environment

Pollution is a by-product of the operation of the power plants. Coal-fired power plants spew out ash and stack gases, internal combustion engines emit exhaust gases, and nuclear reactors leak small amounts of radioactive materials. Besides, all power plants release large amounts of waste heat, which constitutes thermal pollution. What is most noticeable about all this pollution are the immediate, local effects.

The combustion of fossil fuels releases millions of tons of pollutants into the air each year. Carbon monoxide, nitrogen oxides, volatile organic compounds, sulphur dioxide and particulates are commonly referred as "criteria pollutants" because of their contribution to the formation of urban smog. These, also, have an impact on global climate, although, their impact is limited because their radiative effects are indirect, since they do not directly act as greenhouse gases, but react with other chemical compounds in the atmosphere. In particular, the sulphur oxides and nitrogen oxides react with oxygen and water vapor in air to form a fine mist of sulphuric acid and of nitric acid, which is ultimately washed out of the atmosphere in the form of acid rain. The combustion of fossil fuels, such as coal, HFO, diesel and natural gas, liberate three of the major air pollutants, such as, sulphur dioxide (SO_2), nitrogen oxides (NO_X) and particulates (or dust), generally referred in electricity industry as primary emissions.

Particulates can be removed satisfactorily by electrostatic precipitators or cyclones, whereas, the nitrogen oxides emissions can be easily reduced by the use of low NO_X burners. Sulphur dioxide emissions can be reduced by the removal of sulphur from the fuel before combustion, by the removal of sulphur dioxide during the combustion process or by the removal of sulphur dioxide from the flue gases after combustion. The pre-combustion controls comprise selection of low sulphur fuels and fuel desulphurization. The combustion controls are mainly for conventional coal-fired plants and involve in-furnace injection sorbents. The post-combustion controls are the Flue Gas Desulphurization (FGD) processes.

1.8.1. Primary Emissions

1.8.1.1. Sulphur Dioxide Emissions

Sulphur dioxide enters the air mainly from industrial processes and from the combustion of hydrocarbon fuels with a substantial sulphur (S) content and represents a major source of air pollution. Sulphur dioxide is a colourless, corrosive gas that has a bitter taste, but no smell at low levels. The share of SO_2 emissions comes from the use of coal and oil in fossil fired power plants, in industrial combustion units, in small combustion units in households and in vehicles. Although, the concentration of SO_2 in stack gases emitted by steam generation plants is usually in the range of 800 to 6000mg/Nm^3 the volume of gases produced by the utility industry world-wide results in the liberation of large tonnage of sulphur dioxide into the atmosphere.

Sulphur dioxide is a chemical that can be dangerous in many ways. It is known to be lethal to humans at dose levels higher than 1000mg/Nm^3 for a period of over ten minutes. At lower levels sulphur dioxide has been found to be a corrosive irritant to eyes and skin. Sulphur dioxide has been associated with a variety of respiratory diseases and increased mortality rates. Inhalation of sulphur dioxide can cause increased airway resistance by constricting lung passages. Sulphur dioxide is also one of the main ingredients in acid rain. Acid rain occurs when sulphur dioxide or other gaseous chemicals, such as nitrogen oxides, are released into the air. These gases ascend through the atmosphere and when they reach upper cloud levels, they react with water, oxygen and sunlight. Various concentrations of sulphuric acid and nitric acid are then produced. These acids mix with the condensed water vapor in the clouds and fall to the ground with the water in various forms of precipitation. This precipitation with greater acidity is what is known as acid rain. Acid rain is dangerous to many natural ecosystems, it can harm plant, sand, make water undrinkable, and unliveable for many animals.

1.8.1.2. Nitrogen Oxides Emissions

The combustion of HFO results in the formation of various nitrogen oxides emissions, which are generally termed as NO_X. The two most important are nitrogen oxide (NO) and nitrogen dioxide (NO_2). Nitrogen oxide is a colourless gas and less poisonous than the yellowish-brown nitrogen dioxide. The formation of NO increases with temperature, while the proportion of NO_2 decreases. This means that most NO_X formed in a combustion zone consists of NO, which oxidises to NO_2 in the atmosphere. Nitrous oxide (N_2O) is also present in the combustion process in small quantities in conventional combustion boilers.

The formation of NO_X is both due to fuel nitrogen (fuel NO_X) and the oxidation of nitrogen present in the combustion air (thermal NO_X). Generally, there is little thermal NO_X below flame temperatures of 1540°C. On the other hand, fuel nitrogen reacts at a much lower temperature. Control of NO_X emission techniques are based on the chemistry, engineering principles, fluid dynamics and geometry of the gas mixing.

1.8.1.3. Dust Emissions

Dust (or particulates) are the tiny solid or liquid particles that are suspended in air and which are usually individually invisible to the naked eye. Collectively, however, small particles often form a haze that restricts visibility.

1.8.2. Carbon Dioxide Emissions

It is known that CO_2, as one of the products of fossil fuel combustion, is particularly harmful to the earth's climate. The threat to the climate is a global problem, and the topicality of the issue is inextricably bound with the nature of energy production and consumption. The adoption of the United Nations Framework Convention on Climate Change by a significant proportion of States reaffirms the international community's awareness of the problem and its unanimous desire to avert potential global disasters. The States parties to the Convention have committed themselves to stabilizing CO_2 emissions by the year 2000 at levels not exceeding those of 1990. The first major United Nations Conference on Environment was held in Rio 1992, the so-called Earth Summit. Industrialized countries agreed to bring their emissions of greenhouse gases back to 1990 levels by the year 2000. To date, more than 160 Nations are parties to the Climate Convention. The Climate Convection has given rise to the Kyoto Protocol, which was agreed in December 1997, after the Conference of Parties has stated in Berlin in 1995 that the commitment taken so far had not been sufficient. The Kyoto Protocol legally binds the industrialized countries to quantify emission limitation and reductions in the timeframe from 2008 until 2012.

1.8.3. Environmental Legislation

The air pollution standards or allowable emission levels are normally set on two bases such as pollutant concentration from the source of pollution and ambient pollution concentration from all sources. Therefore, the emissions discharge from a smokestack can be controlled by specifying either an allowable discharge concentration level at the stack or a maximum allowable level in the particular locale.

1.8.4. Pricing Environmental Pollution

Environmental cost will play an increasing role in shaping future energy policy. Cleaning up the business of power supply has been declared a fundamental aim of governments around the world, since electricity industry is a major contributor. The task facing energy policy makers is how best to go about the job of reducing pollution in electricity generation when in most countries environmental costs are not reflected in the market price of the end product.

Environmental costs, or at least some elements, are difficult to quantify, but they are nonetheless real. Although there is general agreement as to the broad definition of environmental costs there are widespread variations in defining the boundaries. Arguments for the environmental costs attributable to oil fired generation range from supertanker spillages, to a substantial proportion of western defense budgets. In other words, environmental costs for energy are infernally complicated. For simplification they can, however, be divided into three broad categories: (a) hidden costs borne by governments, (b) costs of the damage caused to health and the environment by emissions other than carbon dioxide and (c) the costs of global warming attributable to carbon dioxide.

The first category includes the cost of regulatory bodies, pollution inspectorates, energy industry subsidies and research and development programmes. The second category, costs due to emissions which cause damage to the environment or create health problems, make up 10-20% of the environmental cost of power generation, depending on the fuel used. They include a wide variety of environmental effects, including damage from acid rain and damage to health from sulphur oxides and nitrogen oxides from thermal power stations. The cost of damage to health is estimated by calculating the loss of earnings and cost of hospitalization of people susceptible to respiratory diseases. Other costs included in the damage and health category are power industry accidents, whether they occur in coalmines, on offshore oil or gas rigs, or in nuclear plant. The probability of a nuclear accident in Western Europe might be extremely low, but should a catastrophic failure occur the costs would be undeniably huge. The third category is by far the largest. Environmental costs due to greenhouse gas emissions which cause global warning, with all its associated effects. This category accounts for some 40-80% of the hidden costs of the world's consumption of electricity. It is, also, the most contentious area of the environmental costs debate. The range of estimates for the possible economic implications of global warning is huge in the studies conducted to date. Costs associated with climate changes, flooding, changes in agricultural patterns and other effects all need to be taken into account.

Clearly, discussion of the make-up of environmental costs is endless. What is important is that governments and regulators should not allow the uncertainty surrounding environmental costs to inhibit acknowledgment of their existence when energy options are being considered. To aid decision makers, studies over the past eight years have identified and quantified the pollution caused by the electricity business. Importantly, they have also tackled the job of pricing that pollution through the so-called "monetarization" of environmental costs.

Many studies have looked at overall damage potentials on health and the environment, assigning a cost penalty to each generating technology, depending on the fuel used (see Table 7). This approach enables the difference in environmental costs between, say, coal and natural gas, to be easily compared. Another approach to monetarization, from a slightly different angle, calculates costs per pollutant. Typically these penalties, as proposed by some states in USA, are around US$10/t for carbon dioxide and up to US$25.000/t of sulphur dioxide. Since the quantities of each pollutant emitted are well known for the differing fuels, these costs enable calculation of the environmental cost of each unit of electricity. Coal fired generation produces the most pollutants, about one kilogram of carbon dioxide per kWh, plus sulphur dioxide and other pollutants, thus attracting the highest penalties.

Assigning cost penalties to polluting technologies brings market forces into play. The price of dirty technology is pushed up and other clean technologies become relatively cheaper. Several USA states have already taken the step of assigning environmental costs to the principal pollutants. The EU approach on greenhouse gas emissions trading would alter the balance between coal, gas, nuclear and renewables, with the proposed tax levels clearly reflecting the mid range of estimates for damage shown in Table 7.

Nevertheless, several European countries have gone ahead and introduced carbon taxes, including Denmark, the Netherlands, Sweden, Finland and Norway. As one of the first, Denmark has implemented a wide range of measures aimed at encouraging its energy industries to become more efficient and to reduce harmful emissions.

Table 7. Environmental costs estimates

Category	Technology			
	Coal	HFO	Natural gas	Nuclear
	USc/kWh			
Human health, accidents	0.70-4.00	0.70-4.80	0.10-0.20	0.03
Crops, forestry	0.17-1.50	1.60-1.70	0.08-0.09	small
Buldings	0.15-5.00	0.20-5.00	0.05-0.18	small
Dissasters	-	-	-	0.11-2.50
Global warming	0.50-24.00	0.50-1.30	0.30-0.70	0.02
Indicative totals	2.05-34.50	3.00-12.80	0.53-1.17	0.16-2.55

The taxes are applied to fuels and to electricity, but also focus on consumption, with space heating, for example, attracting the highest tax rates. However, the tax revenue will be used to encourage measures such as recycling and energy efficiency. In addition, firms will be allowed rebates if they initiate such schemes. The aim is a 20% reduction in carbon dioxide emissions by 2005, one of the highest targets in the world, albeit from a country with an electricity industry emitting more pollution per head of population than almost anywhere in the world.

In the course of the ongoing deregulation of the energy sector, existing energy policies and practices that encourage more environmentally benign energy resources and technologies are being abandoned. New policies are called for in order to secure the gains achieved in the past and to make the further progress needed for sustainability. Carbon taxes can help fill this void. Through raising energy prices, taxes on carbon emissions from fossil-fuel combustion would discourage energy use and induce some switching to less carbon-intensive fuels.

While traditional taxes are decreased, the introduction of pollution taxes would not only decrease pollution, but also provide the revenues needed to sustain the legitimate purposes of a tax system. In short, carbon tax reduces taxes on "goods" (work effort, business activity) and increases taxes on "bads" (pollution, energy use). It offers the promise of a simultaneous improvement of economy and environment. But carbon tax is not a panacea for all social ills. Its scope is constrained by the societal objectives of a tax system and its functional requirements, which pollution taxes alone cannot fulfill. These include equity (whose main vehicle is a progressive personal income tax), and a reliable and steady flow of revenues, which attenuates the impacts of the business cycle. Thus, a significant portion of the current tax system must be maintained. Moreover, as carbon tax can introduce an additional source of revenues, it invites discussion of the uses of those revenues, in addition to tax reductions, such as for technology development and diffusion, training and skills development, education, and infrastructure. The behavioral responses to a carbon tax, based on relatively modest price elasticities of demand, could be amplified by the use of some of its revenues to support efficiency improvements, infrastructure, and market transformation. Thus, while carbon tax appears a highly promising policy strategy, it cannot solely be relied on for environmental policy and should be complemented by other measures, such as, regulation, set-asides, standards, public investments, etc., which can add to and amplify its environmental benefits, and by policies to enhance economic productivity.

The revenues from carbon taxes could provide the means for simplifying existing economic-development incentives. Carbon taxes with offsets in traditional taxes, as the core of a reformed economic-development policy, could give a clear and consistent signal to businesses. Some countries are already engaged in policies to foster the greening of businesses. Carbon tax could enhance this policy goal, wedding environmental and economic policy objectives.

The net effect of a generally more favorable business climate is likely to be positive. Some energy-intensive businesses might very well benefit from the technological advancements that higher energy prices would likely trigger. It is common that energy-efficiency investments improve overall productivity and thus lower not only energy-related costs. Indeed, some countries already maintain economic-development programs that seek to enhance general business productivity through counseling in eco-efficiency (which encompasses energy efficiency and pollution prevention).

A carbon tax would increase the cost of fossil fuels and electricity and generate tax revenues. Electricity producers, industrial and commercial enterprises, and households would respond to these higher fuel prices in three ways: (a) by reducing energy-using services and activities, (b) by investing in energy saving technologies, and (c) by switching to less carbon- intensive fuels. The magnitude of these responses are a function of how easily consumers can change their energy-use patterns, the cost and availability of more energy-efficient equipment and appliances, and the ability of consumers to switch to less carbon-intensive fuels and the technologies that can use them.

For example, in the electric sector, these could be more judicious use of lighting, switching from incandescent to compact fluorescent bulbs, and a shift from coal and oil-fired generation to gas, wind and other low-carbon-intensity technologies. For the commercial and residential sectors, the impacts could be improved thermostat controls, tighter building shells, and a shift from oil to gas. For industry, these could be output changes, process-technology changes (e.g., co-generation), and shifts to gas or biomass. Industry could also respond by a reduction in electricity demand through the use of more efficient lighting, motors, and other equipment in buildings and industry.

The biggest concern voiced about a policy of carbon taxes with other tax reductions is that such a policy would constitute a disadvantage to in-country businesses versus competitors in other countries. This is a concern for a small number of energy-intensive industries, and needs to be addressed through special mechanisms designed to attenuate the negative impacts of carbon tax on these industries. Notwithstanding this concern, cost effects suggest that many industries would indeed experience a net cost decrease from the policies. To the extent that trade flows respond to cost differentials, this would suggest that industries could gain market share in response to the tax policies. The more important point, beyond the scope of the economic analysis, is that carbon tax has the potential to give direction to the technological changes that the economy is undergoing, by steering it onto a path of less carbon-intensive, clean, and future-oriented methods of production and consumption. It is also important that the design of a specific carbon tax take account of a range of equity issues. These include maintaining the society's commitments to progressivity in the modified tax system, and transitional assistance to workers and communities.

It is evident that a carbon tax would likely help meet national climate policy goals, would contribute to significantly improved local and national environmental quality and public health, through a reduction in pollution from energy generation, would provide the opportunity for a consistent economic development policy by yielding revenues to modernize infrastructure and support other economic development policy objectives, such as improving labor skills; and foster technological progress and the development of niche markets.

1.9. Alternative Energy Sources

To generate electricity in developed countries, 70-90% of their energy comes from burning fossil fuels, such as coal, oil and gas, or nuclear power. Fossil fuels release the powerful greenhouse gas carbon dioxide. Climate change, acid rain, mining, oil spills, and radioactive contamination are all unwanted impacts from the unsustainable way in which energy is generate and utilized. As a result of the global warming effect illustrated in Figure 8, recent concerns on environmental protection and sustainable development resulted to the critical need for a cleaner energy technology. Some potential solutions have evolved including energy conservation through improved energy efficiency, a reduction in the fossil fuels and an increase in the supply of environmentally friendly alternative energy sources which is leading to the use of Renewable Energy Sources (RES) and an alternative to large scale source of energy production known as the Distributed Generation (DG) technologies.

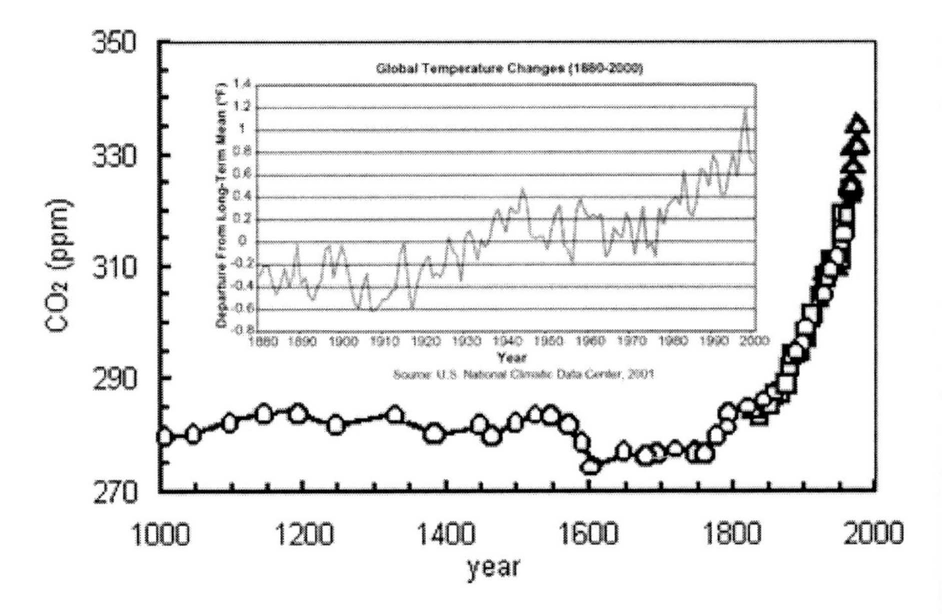

Figure 8. The effect of global warming (Source: U.S. National Climatic Data Center, 2001).

RES and DG technologies are also often called alternative energy sources that use domestic resources that have the potential to provide energy services with zero or almost zero emissions of both air pollutants and greenhouse gases. Renewable energy sources tap naturally occurring flows of energy to produce electricity, fuel, heat, or a combination of these energy types. These inexhaustible sources of energy are domestically abundant and have less impact on the environment than conventional sources. They can provide a reliable source of energy at a stable price. RES technologies are defined as the non-fossil energy sources such as

- wind,
- solar,
- geothermal,
- wave,
- tidal,
- hydropower,
- biomass,
- landfill gas,
- sewage treatment plant gas and
- biogas.

The status of the various RES technologies is indicated in Table 8. An investment in renewable energy (and in energy efficiency) is important to reduce negative economic, social, and environmental impacts of energy production and consumption. As illustrated in Figure 9, RES technologies are already important factors in the world's energy mix, contributing about 18% of the total world electricity requirements. The technical potential of RES technologies is given in Figure 10. The environmental advantages of the production and use of RES electricity seem to be clear. Using RES technologies, like hydropower, biomass, geothermal, and wind energy, in electricity production reduces primary and greenhouse gas emissions.

Table 8. Commercially available and emerging RES technologies

RES technology	Commercially available	Emerging technology
Wind turbines	√	
Photovoltaic systems	√	
Hydropower	√	
Biomass	√	
Geothermal	√	
Fuel cells	√	√
Solar thermal	√	√
Wave energy		√
Tidal energy		√

Emissions such as SO_2, CO_2, and NO_X are reduced considerably, and the production and use of RES electricity contributes to diminishing acid rain and greenhouse effect. Solar- and wind-generated electricity can be used to electrolyze water for producing hydrogen gas.

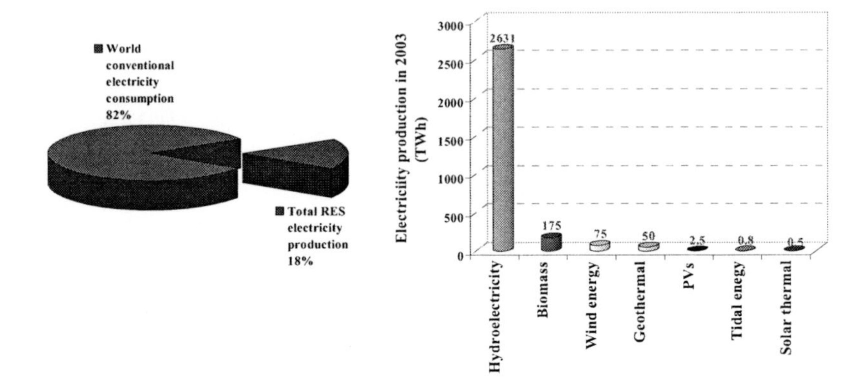

Figure 9. World electricity production in 2003 (Source: EC, 2005).

Hydrogen can be used in fuel cells, which are a promising nonpolluting technology for use as a source of heat and electricity for buildings and as an electrical power source for electric vehicles.

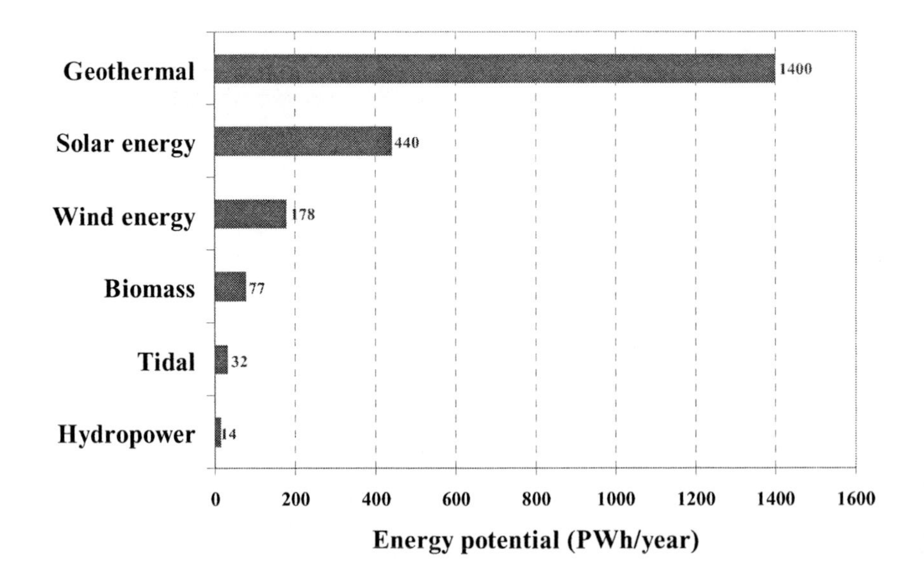

Figure 10. World RES technical potential (Source: World Energy Assessment - UNDP, 2000).

1.10. Future Sustainable Energy Systems

The power industry is one of the few industrial sectors, which affect prosperity of every sphere of economic and social life and exert a direct influence on general technological progress. Much of the world's energy, however, is currently produced and consumed in ways that could not be sustained if technology were to remain constant and if overall quantities were to increase substantially. The need to control atmospheric emissions of greenhouse and other gases and substances will increasingly need to be focused on efficiency and sustainability in energy production.

A sound energy policy should encourage a clean and diverse portfolio of domestic energy supplies. Such diversity helps to ensure that future generations will have access to the energy they need. Renewable energy can help provide for our future needs by harnessing abundant, naturally occurring sources of energy, such as the sun, the wind, and biomass. Effectively harnessing these renewable resources requires careful planning and advanced technology. Through improved technology, we can ensure that clean, natural, renewable and alternative energy sources will be used in the future. Renewable and alternative energy sources will not only help diversify world's energy portfolio but they will do so with few adverse environmental impacts.

Alternative energy includes alternative fuels other than gasoline and diesel, the use of traditional energy sources, such as natural gas, in untraditional ways, such as for distributed energy at the point of use through micro-turbines or fuel cells and future energy sources, such as hydrogen and fusion.

The European Union's (EU) main long-term goal in the field of energy is the conversion of the existing EU energy system, which is heavily dependent on fossil fuels, to a sustainable energy system based on differentiated energy sources of higher energy efficiency. This will enable the EU to face the challenges posed by the security of the energy supply and the climate change while, at the same time, increasing the competitiveness of the European energy industries.

Both renewable and alternative energy resources can be produced centrally or on a distributed basis near their point of use. Providing electricity, light, heat, or mechanical energy at the point of use diminishes the need for some transmission lines and pipelines, reducing associated energy delivery losses and increasing energy efficiency. Distributed energy resources may be renewable resources, such as biomass cogeneration or rooftop solar photovoltaic systems on homes, or they may be alternative uses of traditional energy, such as natural gas micro-turbines.

Currently, Figure 11, oil holds an important position in the EU energy system since it is used widely in the industrial, residential and transport sectors.

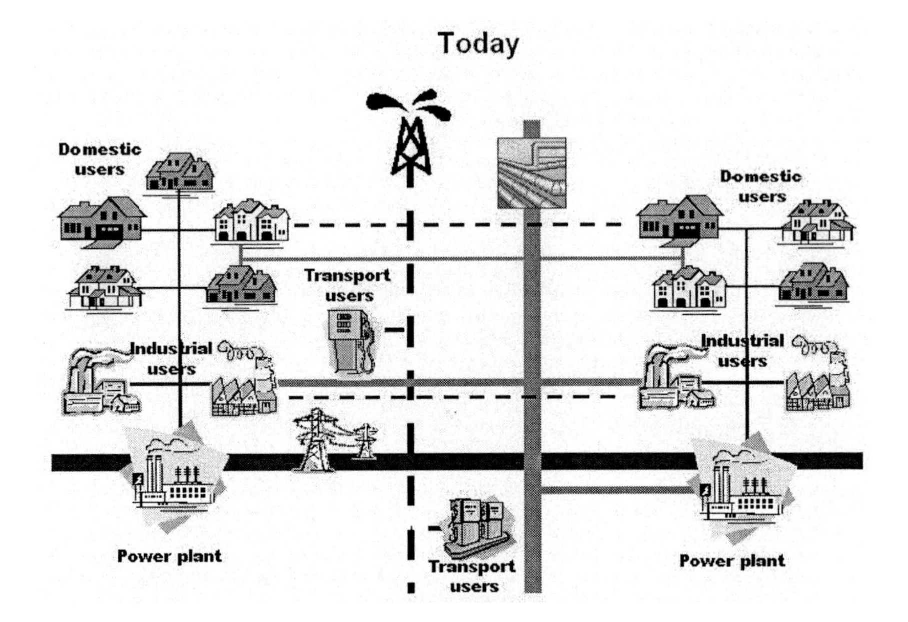

Figure 11. EU energy system today (Source: JRC, 2006).

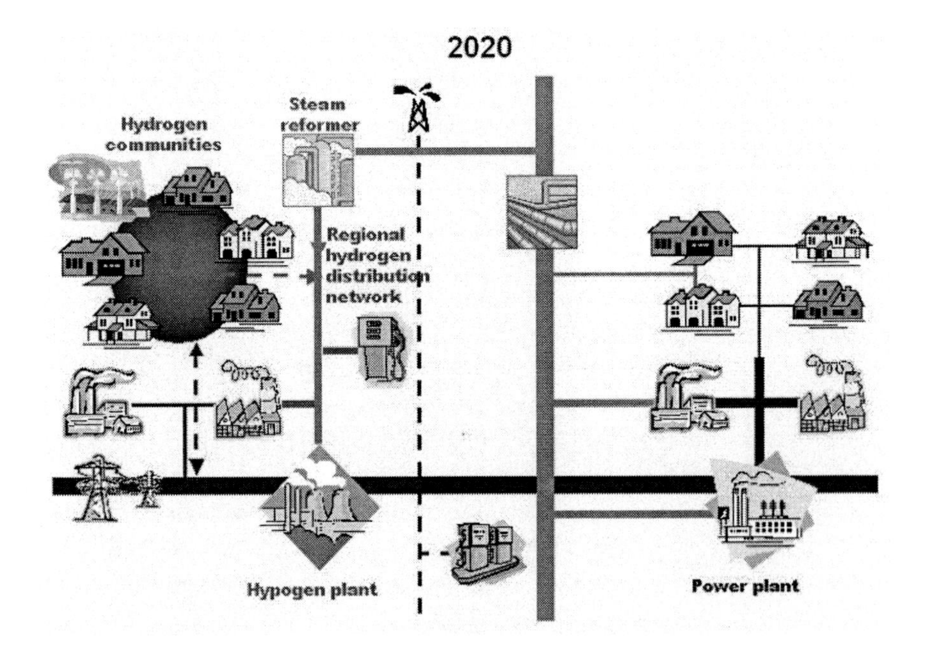

Figure 12. EU energy system in 2020 (Source: JRC, 2006).

Natural gas is also used in all sectors including electricity generation (together with coal and nuclear energy). In the envisaged energy system of 2020, Figure 12, based on the long-term energy targets of the EU, the use of oil will be limited only to the transport sector.

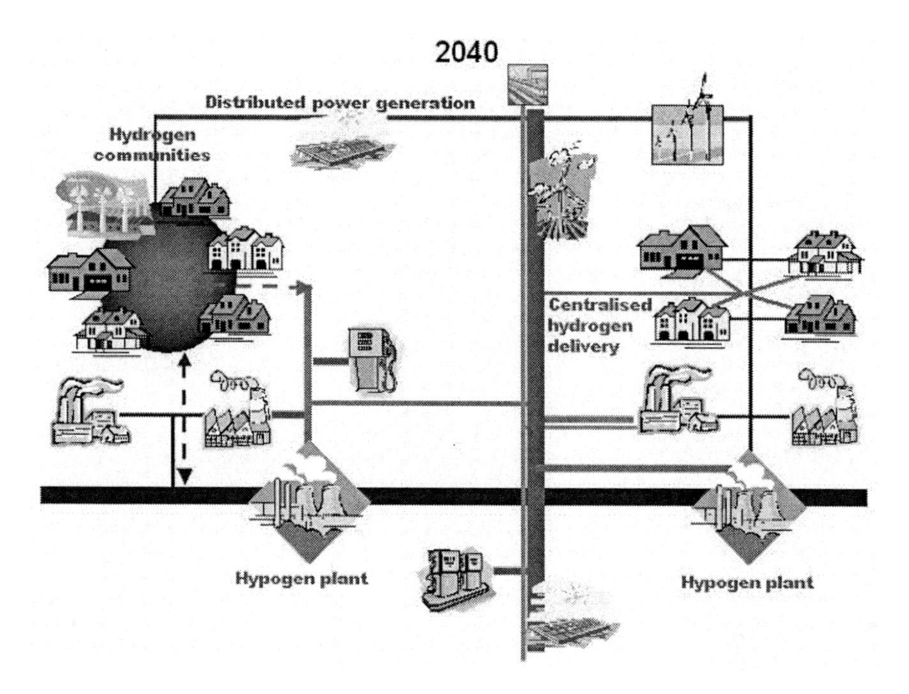

Figure 13. EU energy system in 2040 (Source: JRC, 2006).

Natural gas will emerge as the dominant energy source. In addition, the novel technologies of natural gas reforming and coal gasification with carbon capture and sequestration will, also, be used for hydrogen production which is an environmentally friendly fuel, free of harmful emissions. This will lead, further, to the creation of the first hydrogen communities in which green hydrogen will be produced from renewable energy sources of distributed generation. The first hydrogen power generation units, called HYPOGEN (hydrogen power generation) plants, will be in commercial use and will cover part of the electricity demand without any harmful emissions of carbon dioxide. By 2040, Figure 13, it is hoped that oil will be fully substituted by hydrogen as an energy source. Hydrogen will be generated either by the reforming of natural gas and gasification of coal or by renewable energy sources. This step will complete the transformation of the existing energy economy to a hydrogen economy.

Chapter 2

2. Power Plants

The different types of power generation technologies that can be used for power generation are presented in Figure 14. These are mainly classified into conventional technologies (those technologies using fossil fuels or nuclear fuel for power generation) and renewable technologies. The purpose of this chapter is to present each conventional technology that are currently used for power generation, such as, the Rankine cycle, the simple cycle gas turbine, the combined cycle technologies, etc.

2.1. The Rankine Cycle

The Rankine cycle is the most common cycle used in steam power plants. The Rankine cycle was devised to make use of the characteristics of water as a working fluid and to handle the phase change between liquid and vapor. Also, many other substances can be chosen instead of water as working fluids. The choice depends on many factors, including the need to accommodate the temperatures of heat transfer to and from the vapour and liquid states while maintaining low vapor pressures in the system.

In an idealized simple Rankine cycle shown in Figure 15, heat transfer takes place to the working fluid in a constant-pressure process in the boiler. Liquid enters the pump and is increased in pressure before entering the boiler at relatively low temperature. The energy added to the fluid from the combustion process of the fuel rises the temperature of the fluid producing vapor. Then, the vapor is expanded through a turbine to produce work. In the end of the expansion process, low temperature vapor leaves the steam turbine.

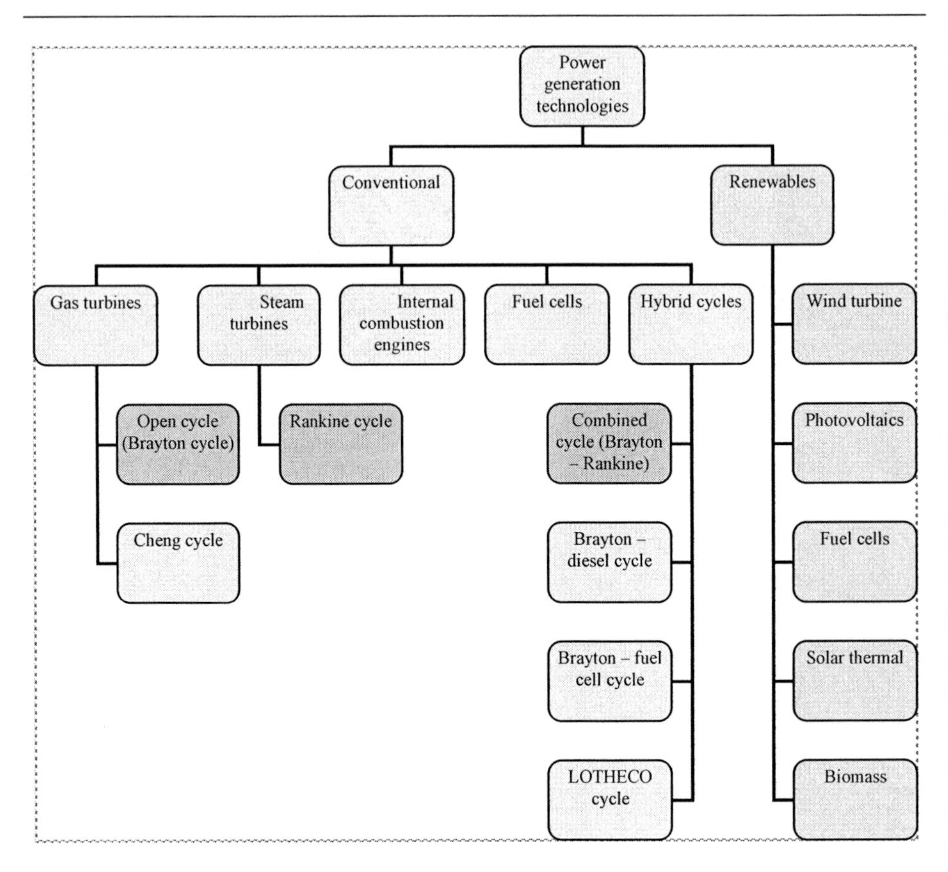

Figure 14. Power generation technologies.

The vapor is then condensed to a liquid as the vapor comes into contact with surfaces in the condenser that are cooled, usually by a cold water stream. Following condensation the liquid enters a pump. The working fluid is returned to the high pressure necessary for energy addition at the higher boiler temperature and the cycle is repeated. Superheating, reheating and regeneration are the most common ways for the improvement of Rankine cycle work ratio and efficiency.

2.1.1. Regenerative Cycle

The regenerative cycle, presented in Figure 16, improves efficiency by the use of feed water heaters to preheat the water before it returns to the steam generator. A portion of the steam is bled from intermediate turbine stages, after it has released part of its energy, and is used to preheat the feed water.

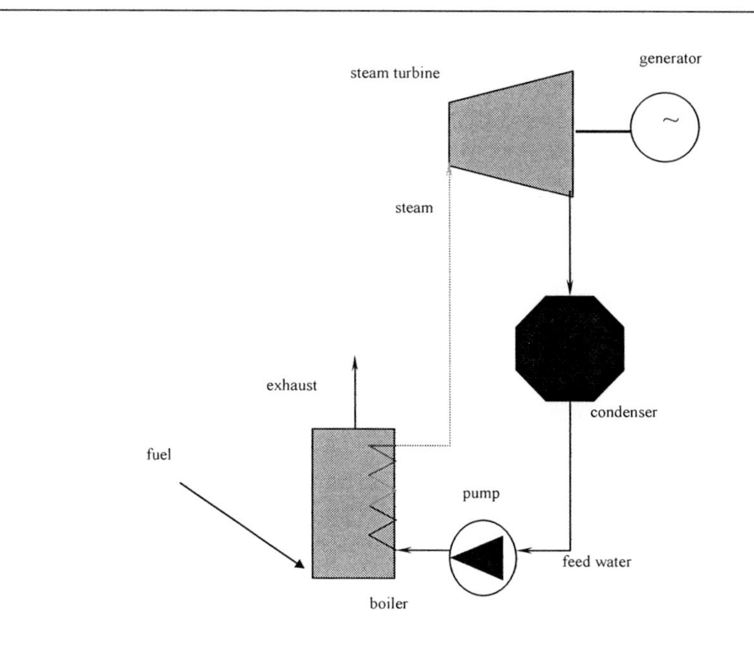

Figure 15. Steam cycle (Rankine cycle).

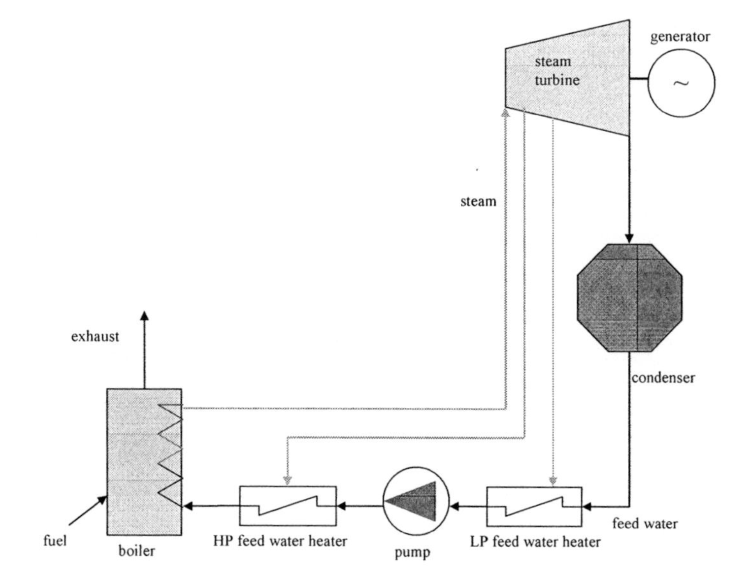

Figure 16. Regenerative Rankine cycle.

The benefit of the process lies in the fact that the latent heat of the bled steam is returned to the main cycle, instead of being rejected in the condenser. Extracting steam before it travels the entire length of the turbine reduces the power output, but this reduction is outweighted by the corresponding efficiency gain. Depending on the particular cycle design details and plant size, several low pressure (LP) and high pressure (HP) heaters are used. The number of heaters and configuration of the heater train is selected to optimize the cycle efficiency, within the appropriate economic constraints.

2.1.2. Reheat Cycle

In the reheat cycle, shown in Figure 17, the steam is partially expanded in the turbine, it is then directed to the reheater where it is superheated again before it returns to the turbine where it continues its expansion. This design increases both the efficiency and power output of the cycle, at the expense of increased initial cost.

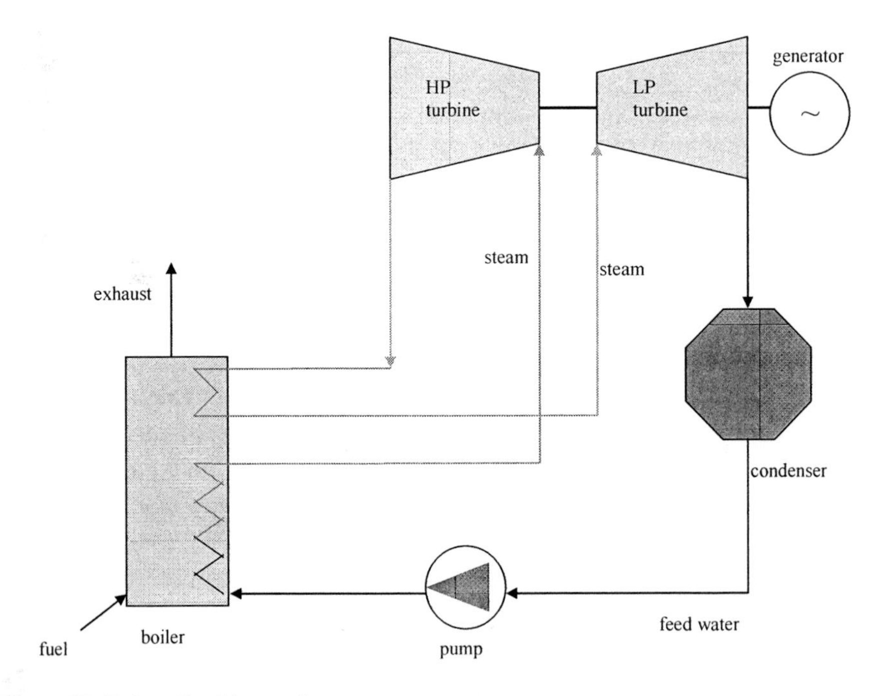

Figure 17. Reheat Rankine cycle.

Typically the turbine is split in two, the HP and LP turbines, to allow better handling of the temperature gradients. The reheat pipework is also another important design consideration, both in terms of engineering and economics. Because the reheat steam is at a lower pressure than the HP steam, larger diameter pipes are required, necessitating increased support, increased expansion allowances and of course increased capital outlay.

2.2. The Simple-Cycle Gas Turbine

A schematic diagram for a simple-cycle gas turbine, for power generation, is shown in Figure 18. Air enters the axial flow compressor at point 1 at ambient conditions. The standard conditions used, since these vary from day to day and from location to location, by the gas turbine industry are 15°C, 1,013bar and 60% relative humidity, which are established by the International Standards Organization (ISO) and referred to as ISO conditions.

Air entering the compressor at point 1 is compressed to some higher pressure. No heat is added, however, compression raises the air temperature so that the air at the discharge of the compressor is at a higher temperature and pressure. Upon leaving the compressor, air enters the combustion chamber at point 2, where fuel is injected and combustion occurs. The combustion process occurs at essentially constant pressure. Although, high local temperatures are reached within the primary combustion zone (approaching stoichiometric conditions), the combustion system is designed to provide mixing, burning, dilution and cooling. Thus, by the time the combustion mixture leaves the combustion system and enters the turbine at point 3, it is at a mixed average temperature. In the turbine section of the gas turbine, the energy of the hot gases is converted into work.

This conversion actually takes place in two steps. In the nozzle section of the turbine, the hot gases are expanded and a portion of the thermal energy is converted into kinetic energy. In the subsequent bucket section of the turbine, a portion of the kinetic energy is transferred to the rotating buckets and converted to work. Some of the work developed by the turbine is used to drive the compressor, and the remainder is available for useful work at the output flange of the gas turbine. Typically, more than 50% of the work developed by the turbine sections is used to power the axial flow compressor. The thermodynamic cycle upon which all gas turbines operate is called the Brayton cycle (or the Joule cycle). Although, the exhaust is released at temperature of 400°C to 600°C and represents appreciable energy loss, modern gas turbines offer high efficiency (up to 42%) and a considerable unit power output (up to 270MWe).

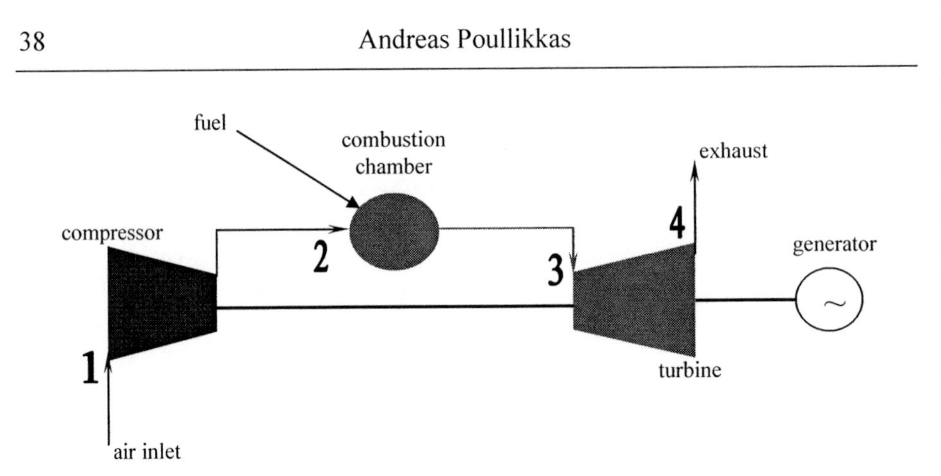

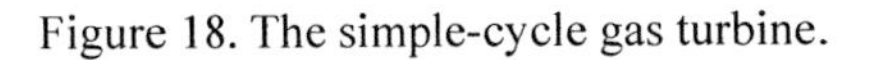
Figure 18. The simple-cycle gas turbine.

One important disadvantage is that a gas turbine does not perform well in part-load operation. For example, at 50% load, the gas turbine achieves around 75% of the full load efficiency, and at 30% load this drops to 50% of the nominal efficiency. Therefore, arrangements, such as the controlled inlet guide vanes and multi-shaft designs, are employed to improve the part-load performance. Other modifications of the cycle include reheat, inter-cooling and recuperation. The expansion work can be increased by means of reheating. Moreover, this makes it possible to provide full-load efficiency within a broader load range by varying reheat fuel flow. Because of the increased specific work output due to reheat, the plant becomes compacter. Another technique to increase the specific work output is inter-cooling, which diminishes the work required by the compressor. The compressor outlet air becomes colder and, if air cooling is applied, this allows higher turbine inlet temperatures.

2.3. The Gas to Gas Recuperation Cycle

Gas turbine efficiency can be raised when gas to gas recuperation is employed and this has been used in conjunction with industrial gas turbines for more than 50 years. This arrangement is illustrated in Figure 19. The use of recuperation is limited, however, by the compressor outlet temperature due to metallurgical problems of the heat exchanger temperature. Inter-cooling reduces the heat transfer problem and allows recuperation with high efficiency turbines.

This concept is used in several gas turbines, such as the 1.4MWe Heron gas turbine, the 21MWe Rolls-Royce WR–21 gas turbine, or the Solar gas turbines in the 1MWe to 25MWe size range. The recuperated gas turbines are expected to obtain efficiencies from 39% to 43%, which are higher compared to 25% to 40% for other simple-cycle gas turbines of same capacity.

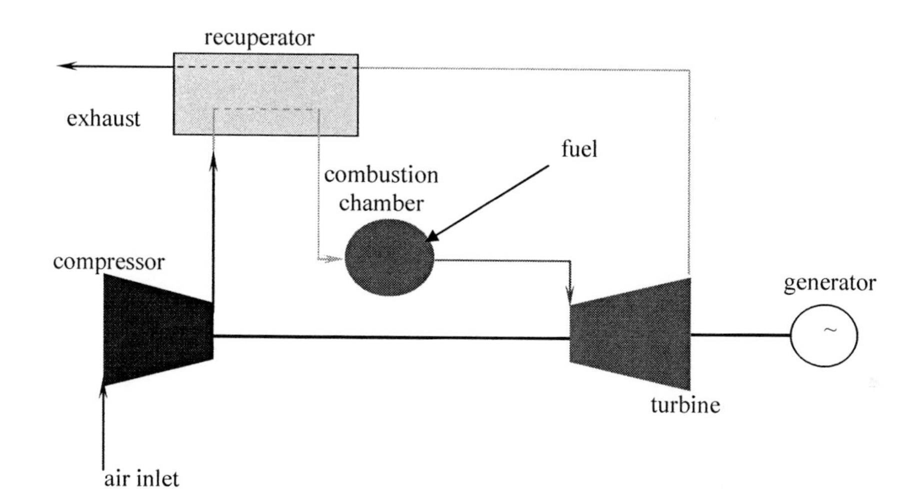

Figure 19. Gas to gas recuperation.

There is a view, which is well supported theoretically, that a regenerator would provide a more efficient cycle than a recuperator. This is due to the very high thermal effectiveness of regenerators since values of around 95% are possible. However, maximum efficiency occurs at very low pressure ratios, typically less than 5. Consequently, even if the theoretical performance could be achieved, relatively large turbines would be required.

2.4. The Combined Cycle

A typical simple-cycle gas turbine will convert 30% to 40% of the fuel input into shaft output. All but 1% to 2% of the remainder is in the form of exhaust heat. The Brayton – Rankine cycle, commonly referred as to the combined cycle is the well-known arrangement of a gas turbine with a steam turbine bottoming cycle. The combined cycle is generally defined as one or more gas turbines with heat recovery steam turbines in the exhaust, producing steam for a steam turbine generator. Figure 20, shows a combined cycle in its simplest form. High utilization of the fuel input to the gas turbine can be achieved with some of the more complex heat recovery cycles, involving multiple-pressure boilers, extraction or topping steam turbines, and avoidance of steam flow to a condenser to preserve the latent heat content. Attaining more than 80% utilization of the fuel input by a combination of electrical power generation and process heat is not unusual. Combined cycles producing only electrical power are in the 50% to 58% thermal efficiency range using the more advanced gas turbines.

In a typical scheme, shown in Figure 20, exhaust heat from the open gas turbine circuit is recovered in a heat recovery steam generator. In order to provide better heat recovery in the heat recovery steam generator, more than one pressure level is used. With a single pressure heat recovery steam generator typically about 30% of the total plant output is generated in the steam turbine. A dual pressure arrangement can increase the power output of the steam cycle by up to 10%, and an additional 3% can result by choosing a triple pressure cycle. Modern gas turbine combined cycle plants with a triple pressure heat recovery steam generator with steam reheat can reach efficiencies above 55%. Siemens/Westinghouse claims 58% efficiency, Alstom claims 58.5% efficiency and General Electric claims an efficiency of 60%. These high efficiency values can be achieved at large units above 300MWe.

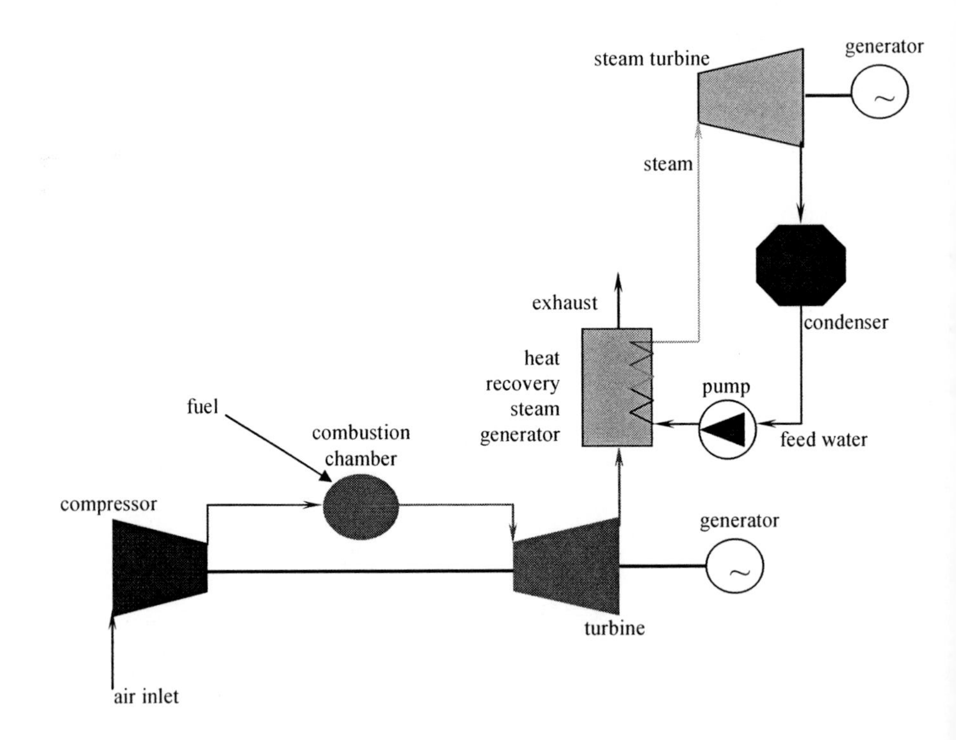

Figure 20. The combined cycle.

Combined cycle plants have become a well-known and substantial technology for power generation due to its numerous advantages including high efficiency and low emissions. The combined cycle technology provides a range of advantages. These include: (a) higher thermal efficiency from any other gas turbine advance cycle (the best conventional oil- and coal-fired power plants on the market have thermal efficiencies around 43%–45%), (b) low emissions since natural gas produces no ash or SO_X, less quantities of volatile hydrocarbons, carbon monoxide and NO_X than oil and coal and produces much less CO_2, (c) low capital costs and short construction times (often 2–3 years), (d) less space requirements than the space required for equivalent coal or nuclear stations, which reduces site constraints, (e) flexibility in plant size with maximum power outputs range between 10MWe and 750MWe per combined cycle –unit and (g) fast start-up which makes it easier to respond to changes in demand.

2.5. The Brayton - Kalina Cycle

The Kalina cycle is a novel bottoming cycle which uses zeotropic mixture of ammonia and water as the working fluid. Its characteristics are such that its temperature tracks the turbine exhaust temperature in the waste heat boiler. However, the thermodynamic advantages of this small boiler temperature difference compared to a steam cycle would be lost at the condensing stage, assuming the condenser cooling medium temperature would be the same in both cases. The novelty of the Kalina cycle lies in the solution to this problem.

The principle of the Brayton - Kalina cycle is illustrated schematically in Figure 21. There are, in effect, two condensing stages. In the first stage, the turbine exhaust stream is fully absorbed, on a continuous basis, by a stream of secondary fluid in a liquid phase and the heat of absorption is dissipated into the condenser cooling water. The secondary fluid is an ammonia/water mixture of different composition from the turbine exit stream but this is not important with respect to the underlying principle. Following the absorption process, the mixture of secondary fluid plus turbine exit stream is pressurised by a pump. Since a liquid is being compressed, very little pumping work is required. The pressurised fluid is heated by the turbine exhaust stream and this causes the turbine working fluid to boil off from the secondary fluid. Since the fluid is now at a higher pressure than at the turbine exit it can be fully condensed by the condenser cooling water.

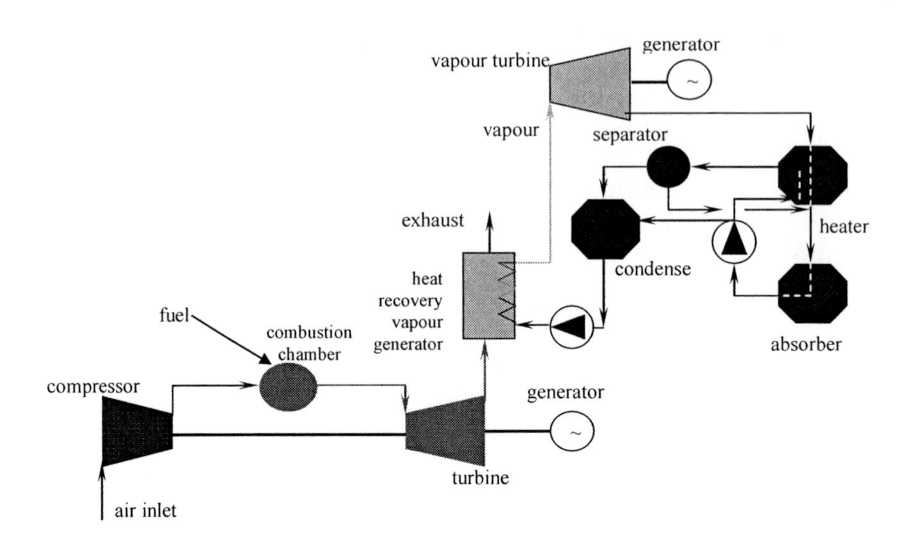

Figure 21. The Brayton - Kalina cycle.

The Kalina cycle can produce 10% to 30% more power than a Rankine cycle. Because the exhaust pressure of the vapour turbine in the Kalina cycle is above atmospheric pressure, no vacuum is needed to be maintained in the condenser during operation, or stand-by periods. Therefore, the start-up procedure can be performed in a much shorter time. The working fluid composition can easily be changed in order to obtain the optimal performance in respect to alterations in load or ambient conditions. Another advantage is the smaller size of the whole unit. The footprint of the Kalina plant is about 60% of the size of a Rankine plant design.

In 1993 General Electric signed an exclusive worldwide licensing agreement with the Kalina cycle patent owner (Exergy, Inc.) to design and market Brayton – Kalina cycle plants. The General Electric commercial size demonstration plant, of 260MW capacity, was planned to be in operation by 1998, however, the project was suspended.

2.6. The Brayton – Brayton Cycle

Two Brayton cycles can be combined by an air-gas heat exchanger as illustrated in Figure 22. The exhaust of the primary gas turbine is sent to a heat exchanger, which, in turn, heats the air in the secondary gas turbine cycle. Air is expanded in the turbine to generate additional power. In comparison to the conventional combined cycle, this scheme does not require bulky steam equipment (boiler, steam turbine, condenser), or a water processing unit, and allows unmanned operation.

Recent studies showed the feasibility of this configuration. These reported an increase of power by 18% to 30% depending on the number of intercoolers, and an efficiency growth of up to 10% points. For example, for the Allison 571K topping gas turbine, introduction of the air bottoming cycle with two intercoolers led to an increase in power from 5.9MWe to 7.5MWe and in efficiency from 33.9% to 43.2%. Comparable results were obtained with the General Electric LM2500 topping turbine. The scheme can also be applied for cogeneration. The exhaust air, leaving the cycle at 200°C–250°C, can be used for process needs that require heat of such temperatures.

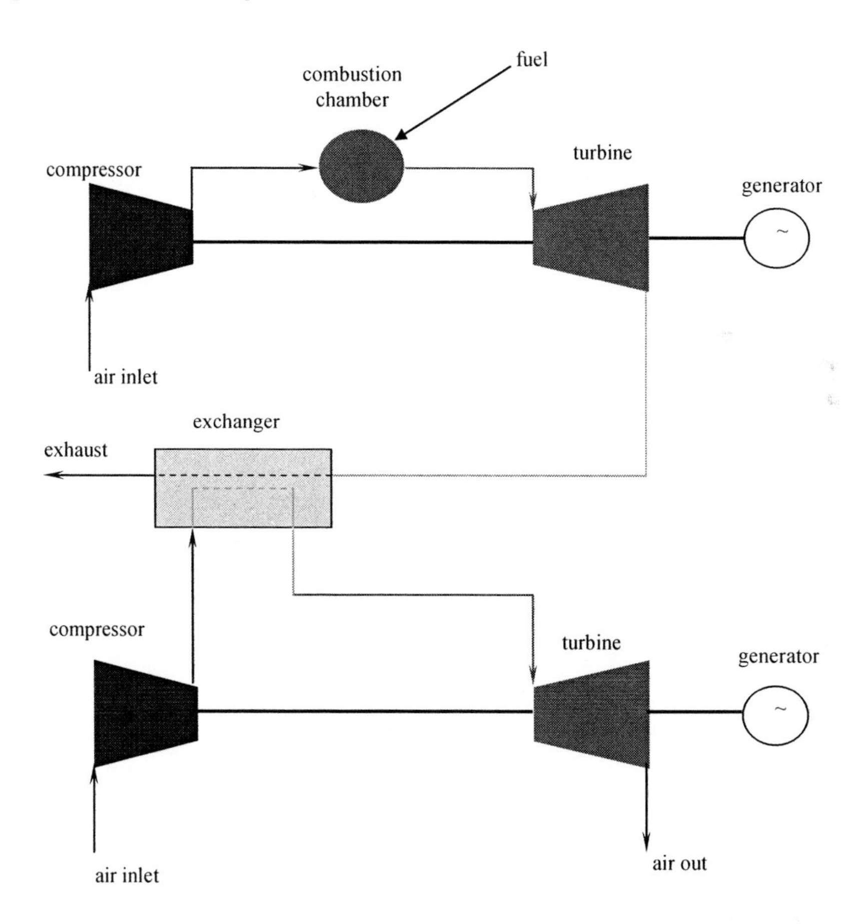

Figure 22. The Brayton - Brayton cycle.

2.7. THE BRAYTON – DIESEL CYCLE

Preheating of the inlet air of a Diesel engine can sufficiently improve its performance. The gas turbine exhaust can be applied in order to increase the temperature of the air, which is extracted from the compressor and fed into the Diesel engine. Subsequently, the engine outlet flow expands through the low-pressure stage of the gas turbine as illustrated in Figure 23.

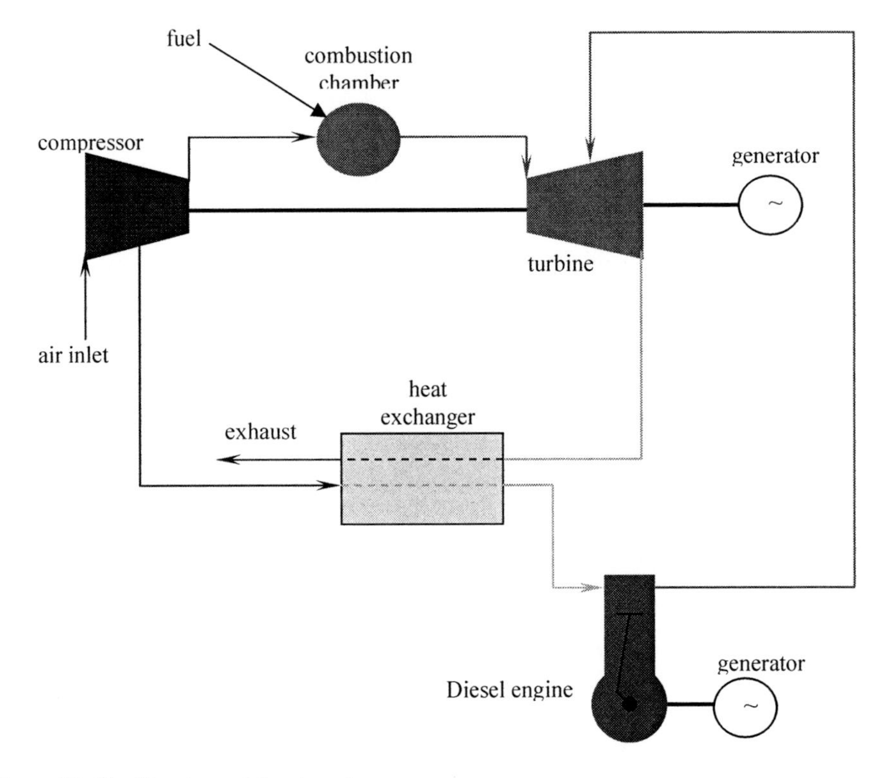

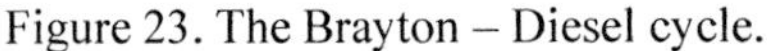
Figure 23. The Brayton – Diesel cycle.

2.8. THE BRAYTON – STIRLING CYCLE

In a combination of a gas turbine and a Stirling engine, the heater of the latter can be placed either in the combustion chamber of the turbine, or after the turbine in the exhaust flow, as shown in Figure 24. The arrangement is determined by the optimal performance of the cycle and by the materials used in the Stirling heater's head. As much as 9MWe can be recovered by a bottoming Stirling cycle from the exhaust of a Rolls-Royce RB211 gas turbine of 27.5MWe. Such a plant can obtain an efficiency of 47.7%. Just as the Brayton - Brayton cycle, this combination provides a compact and simple heat recovery scheme.

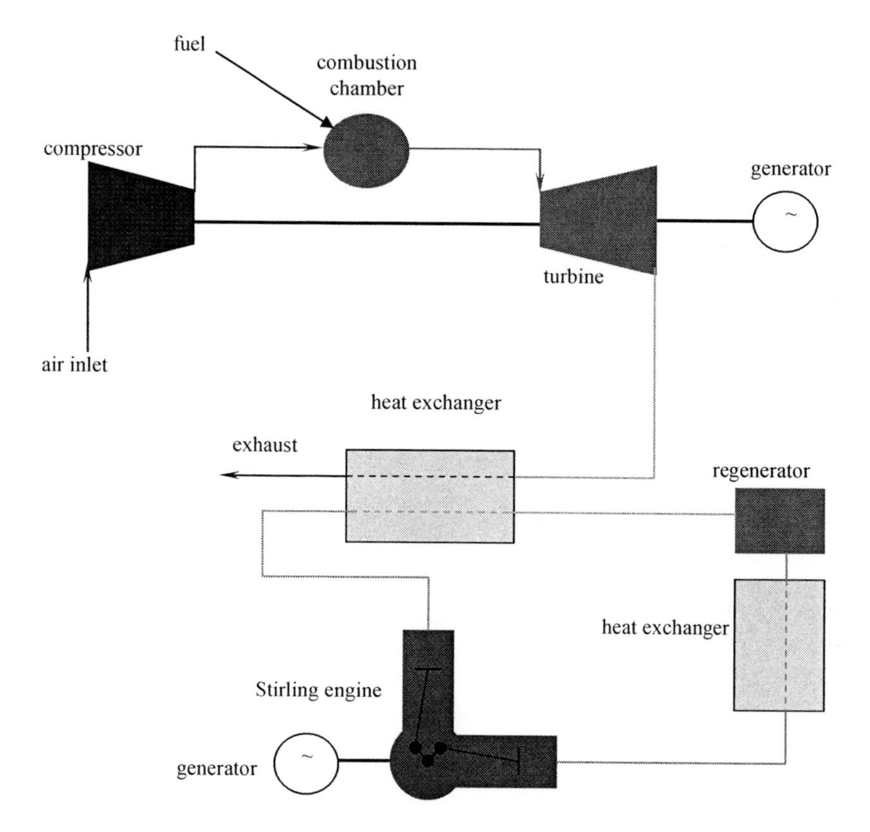

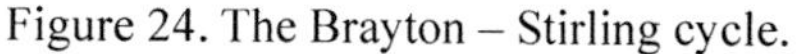
Figure 24. The Brayton – Stirling cycle.

2.9. The Brayton – Fuel Cell Cycle

A fuel cell system (for details see section 3.3.2 The fuel cell technology on page 151), which offers high efficiency (60%), can operate at high pressure and can produce very high temperature exhaust gases, which allows integrating a gas turbine within the system, thus improving performance. The schematic of the system is presented in Figure 25. The use of the fuel cells integrated with combustion chambers allows efficiency to approach 70%. The Brayton – fuel cell cycle is claimed to have the highest efficiency of any advance cycle, and can, therefore, be seen as a choice for the future power plants.

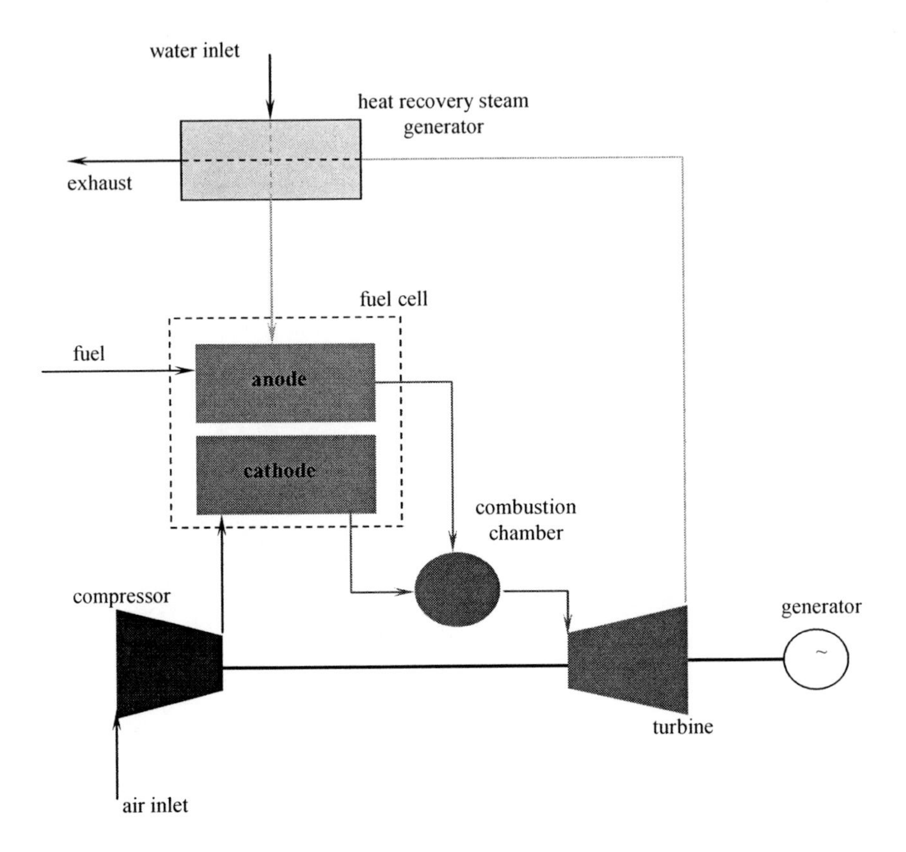

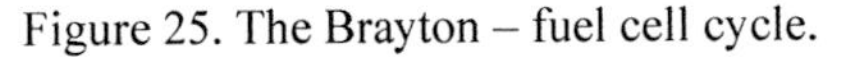
Figure 25. The Brayton – fuel cell cycle.

2.10. The Chemical Recuperation Cycle

The chemical recuperation cycle uses a reforming process to convert methane, water, and sometimes CO_2 into a hydrogen and carbon monoxide fuel mixture that can be burned in the combustor. This endothermic reaction absorbs heat at a temperature lower than the combustion temperature and in this manner increases the fuel's heating value. Recuperation that proceeds thermally and chemically results in a higher degree of heat recovery than in standard recuperation schemes. Moreover, the hydrogen rich fuel has greater flammability than methane and supports combustion at a lower flame temperature, which potentially reduces NO_X formation.

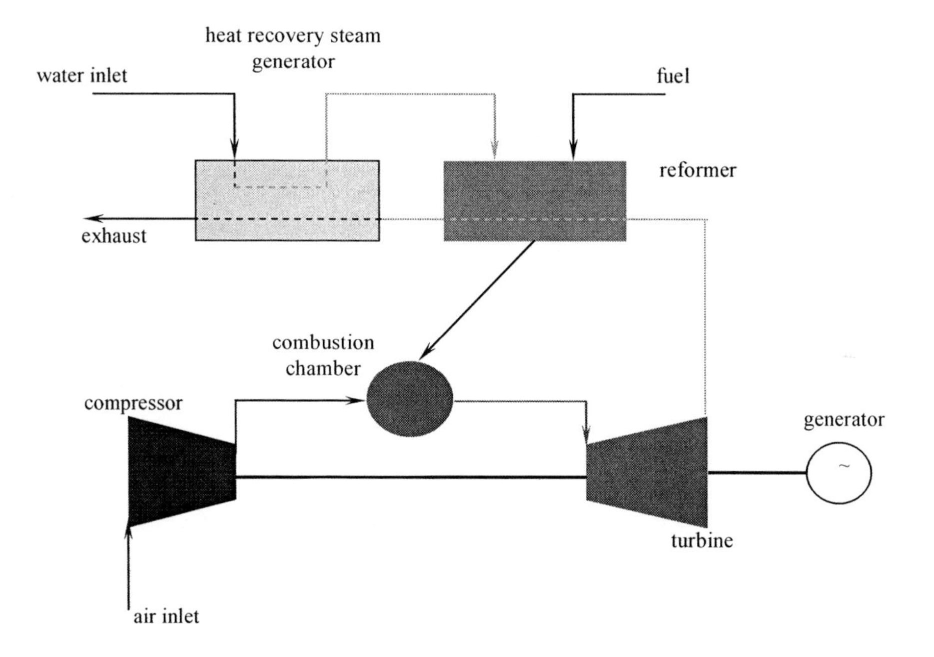

Figure 26. The chemical recuperation cycle with steam reforming.

However, the gas turbine exhaust temperature is not high enough for a complete reforming reaction. At 550°C only 20% of the fuel is reformed. In order to increase the temperature some additional firing can be applied. Different reforming schemes have been proposed. In the scheme with steam reforming, steam generated in a heat recovery steam generator is mixed with natural gas in a reformer as illustrated in Figure 26. A power plant based on the General Electric LM5000PC gas turbine could reach an efficiency of 45% in comparison to 37,2% in simple-cycle.

Another scheme uses exhaust gas recuperation, a portion of the flue gas is compressed, mixed with natural gas, heated with exhaust heat from the combustion turbine, mixed with the air from the compressor, and sent to the combustion chamber (Figure 27). When the mixture is heated in the presence of a nickel-based catalyst, hydrogen and carbon monoxide are produced. The reaction is accelerated at low excess oxygen, low pressure, and high mass ratio of recycled exhaust gas to methane. Therefore, the best results are achieved when the fuel is burned at the stoichiometric ratio, and the exhaust gas is used to reduce the turbine inlet temperature. Typical value of recirculation as over 50% of turbine flow. This means that both the air compressor flow and the exhaust flow are less than half that of conventional cycles with the same turbine size.

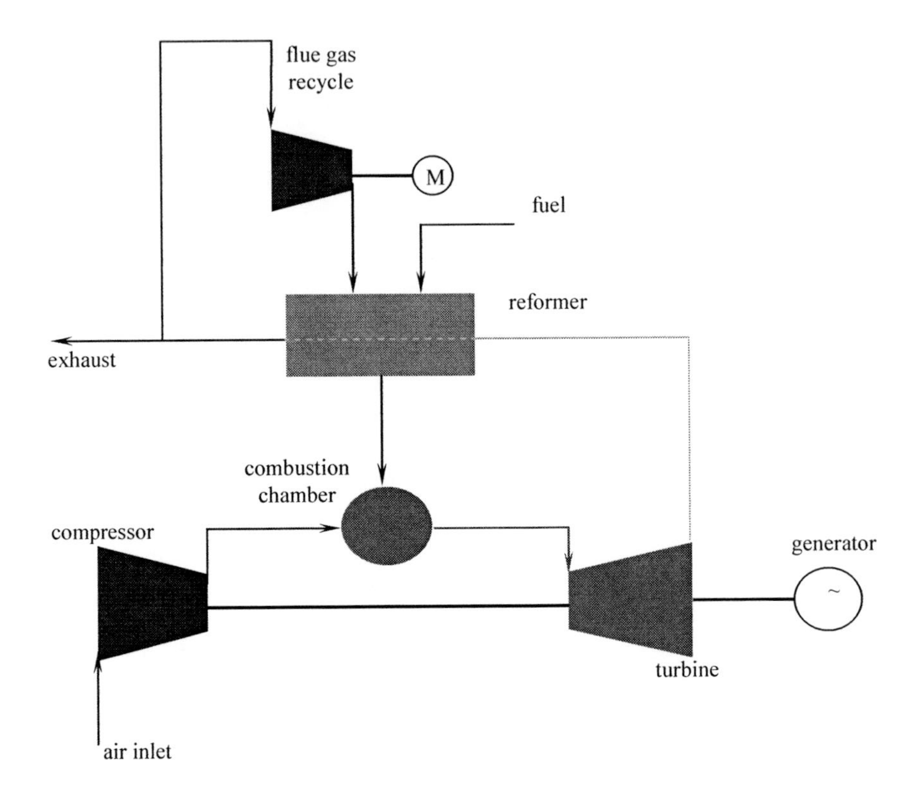

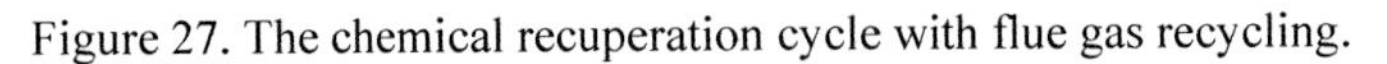
Figure 27. The chemical recuperation cycle with flue gas recycling.

2.11. MAST Cycles

Mixed Air Steam Turbines (MAST, also, referred as wet gas turbines or as mixed gas steam turbines) are an interesting technology for power generation since these can improve the performance of gas turbine based power plants at feasible costs. In particular, MAST technologies can improve the performance of a simple-cycle gas turbine by the integration of a bottoming water/steam cycle into the gas turbine cycle in the form of water or steam injection. Such configuration has a higher electrical efficiency than the simple-cycle gas turbine and produces more electricity per unit fuel input. Well-known schemes of this technology are the steam injection gas turbines and the humid air turbines.

Water injection has been used for power augmentation in aircraft engines since the 1950s, and in industrial gas turbines since the 1960s. The injection increases the mass flow and the specific heat of the working fluid, which gives additional power to the cycle. Along with this, it helps to lower NO_X formation in the combustion chamber and to cool the blades more effectively than air. Depending on the amount of water or steam injection output will increase because of the additional mass flow. Figure 28 shows the effect of steam injection on the power output for a typical gas turbine.

There are five gas turbines on the market, which are adapted to the use of steam injection. The three bigger machines are all constructed by General Electric[1] and are the LM5000 STIG™, the LM2500 STIG™ and the LM1600 STIG™, producing respectively 51.6MWe, 28.1MWe and 17MWe. Without steam injection they produce 34.5MWe, 22.8MWe and 13MWe. The two smaller machines are the Allison 501-KH and the Kawasaki M1A-13CC. The most recent variant of the Allison 501 produces 4.9MWe without steam-injection and 6.8MWe with steam injection.

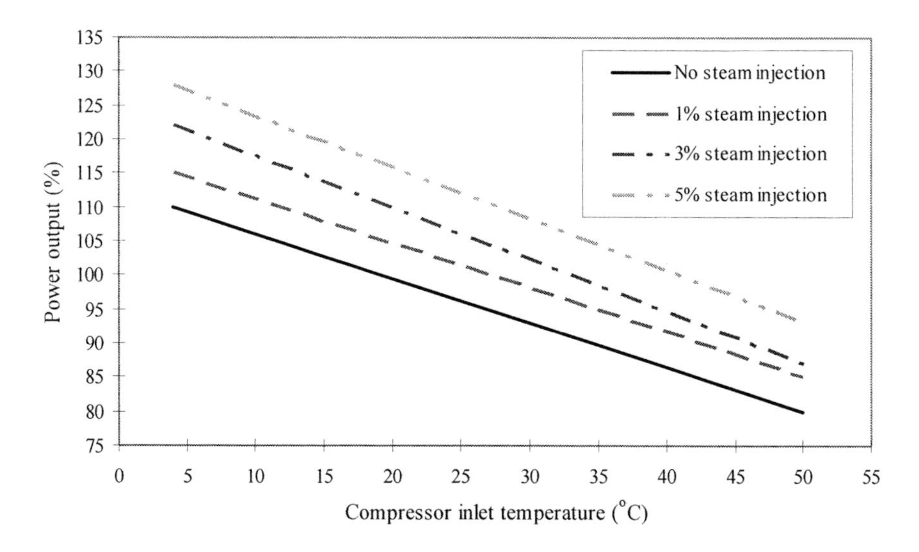

Figure 28. Effect of steam injection on power output.

[1] PT General Electric dubbed its steam injected gas turbines STIG™ (STeam Injected Gas Turbine).

The latest development in steam injected gas turbines is the Kawasaki M1A-13CC. With this machine Kawasaki aims at the low power co-generation applications. The gas turbine produces 2.4MWe in steam injection mode and 1.3MWe without steam injection. Various types of MAST turbines are currently under development.

In the following sections an overview of the MAST technologies is presented. These include the Cheng cycle, the DRIASI cycle, the HAT cycle, the LOTHECO cycle, etc. The main characteristics are given along with a discussion on advantages and disadvantages of these schemes.

2.11.1. The Cheng Cycle

In 1978 Cheng proposed a gas turbine cycle in which the heat of the exhaust gas of the gas turbine is used to produce steam in a heat recovery steam generator as shown in Figure 29.

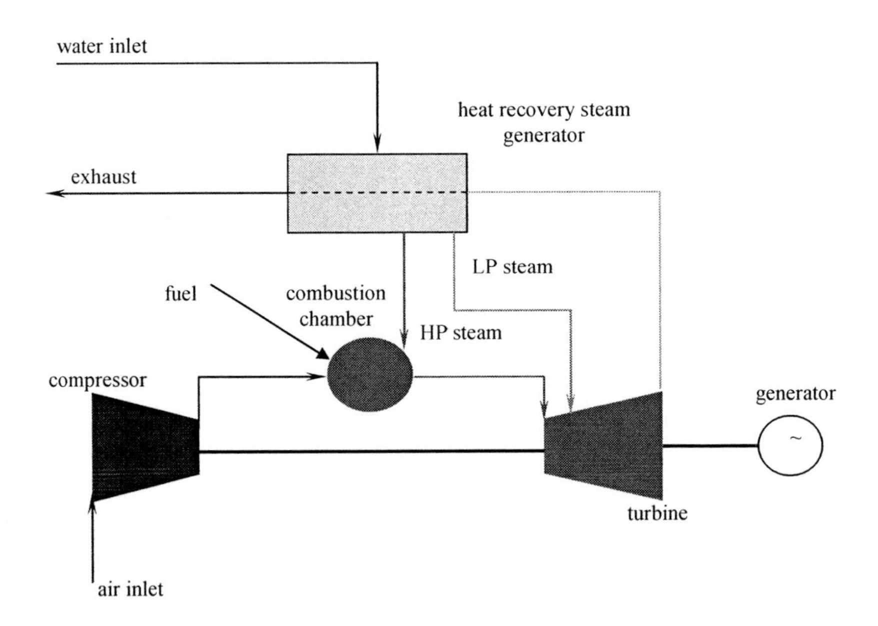

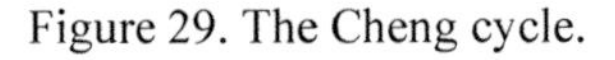
Figure 29. The Cheng cycle.

This steam is injected in the combustion chamber of the gas turbine, resulting in an efficiency gain and a power augmentation. The cycle is commonly called the Cheng cycle or the steam injection cycle. High-pressure steam can be injected into the combustion chamber, while intermediate-pressure and low-pressure steam is often expanded in the first gas turbine stages, as shown in Figure 29.

The system will work if the pressure of the steam is higher than that at the compressor outlet. By introducing steam injection in a gas turbine an efficiency gain of about 10% and a power augmentation of about 50% to 70% are possible. Using a steam turbine to expand the steam, i.e., applying a conventional combined cycle instead of a Cheng cycle, gives higher efficiency gains. Accepted efficiency for the combined cycle is nowadays 50%-58%, with a power rise of about 30% to 50% with respect to the simple cycle.

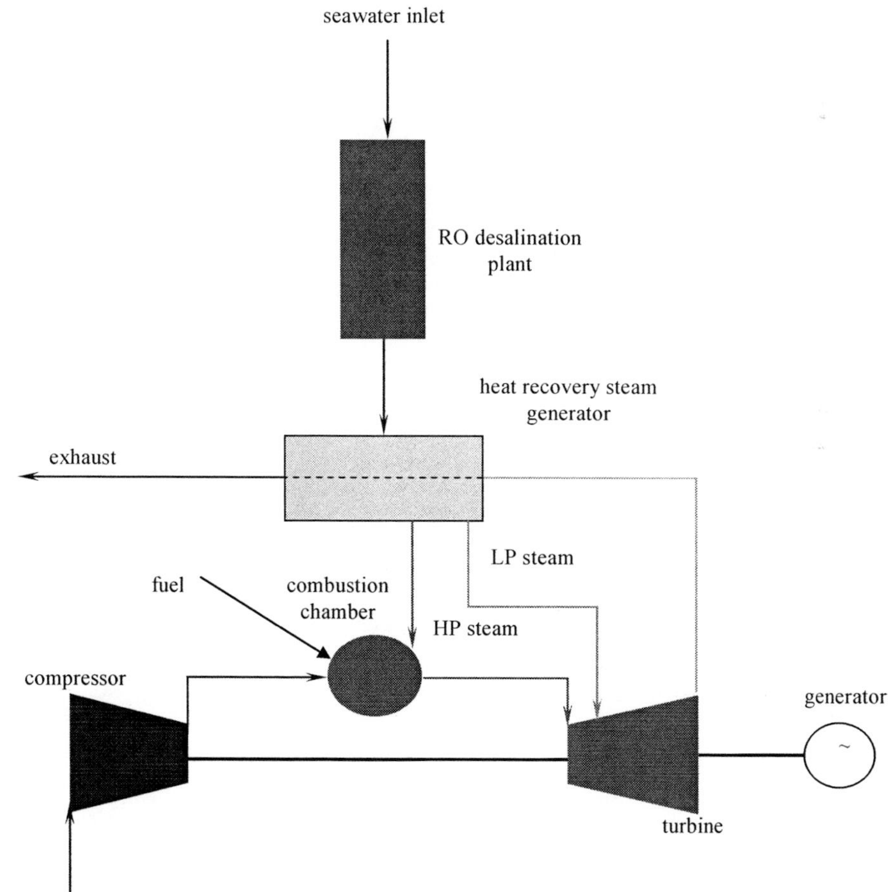

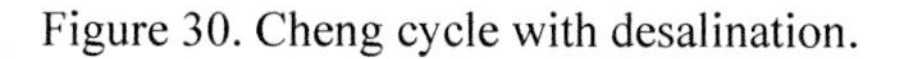
Figure 30. Cheng cycle with desalination.

Expansion of steam in the gas turbine proceeds to the atmospheric pressure and in a less efficient manner than in the steam turbine. Whereas in the combined cycle plant steam leaves the steam turbine at much lower pressures, thus providing more power. Therefore, a gas turbine with steam injection will always have a lower efficiency than that in conventional combined cycle operation.

The introduction of intercooling and reheat in a steam injection gas turbine allows to reduce the power consumed by the compressor from 50% of the total output for modern engines down to 30%. Therefore, efficiency of the gas turbine becomes less dependent on the compressor characteristics and the work ratio is considerably increased.

Because, in modern designs, the temperature of injected steam must be raised in the gas turbine combustion chamber to about 1250°C, there are advantages in a close approach temperature in the waste heat boiler. However, because boiling is a constant temperature process, high steam temperatures are necessarily associated with a low level of heat recovery. Conversely, the steam temperature must be relatively low if the heat recovery from the exhaust stream is to be maximised. This difficulty limits the maximum theoretical efficiency of the overall system. In practice, the high temperature option is normally the best compromise.

A practical concern with steam injection is water consumption. Consumption is, typically, between 1.1kg and 1.6kg of high purity water per kWh of electrical output. The necessary water purification system for large scale plant would represent about 5% of the total capital expenditure and running costs would add about 5% to the fuel cost.

Typical designs with integration into the Cheng cycle a reverse osmosis (RO) desalination plant or a water recovery condenser are presented in Figure 30 and Figure 31 respectively. Recent calculations indicated that both configurations are cost effective for acute areas with water shortages.

2.11.2. The steam injected cycle with topping steam turbine

A steam injected gas turbine cycle with a topping steam turbine is shown in Figure 32. A high pressure steam is first expanded in a back-pressure steam turbine, producing power, and then is injected into the combustion chamber of the gas turbine.

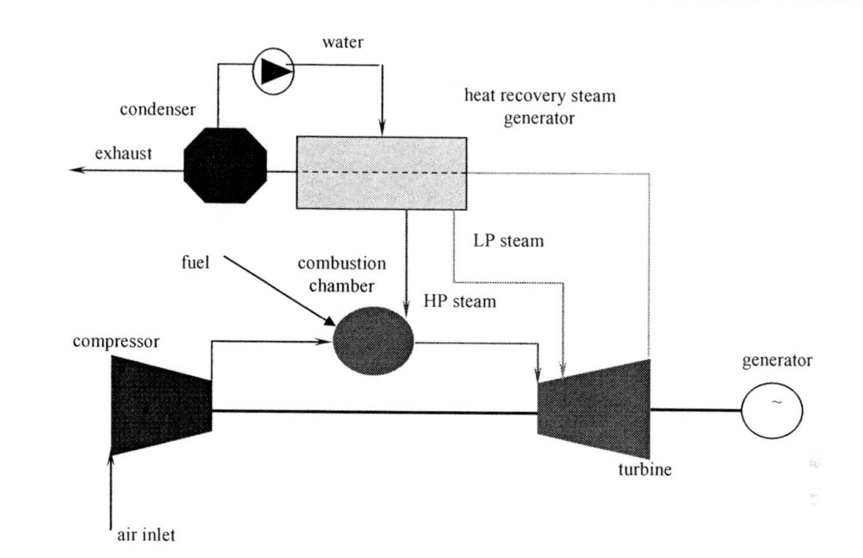

Figure 31. Cheng cycle with water recovery condenser.

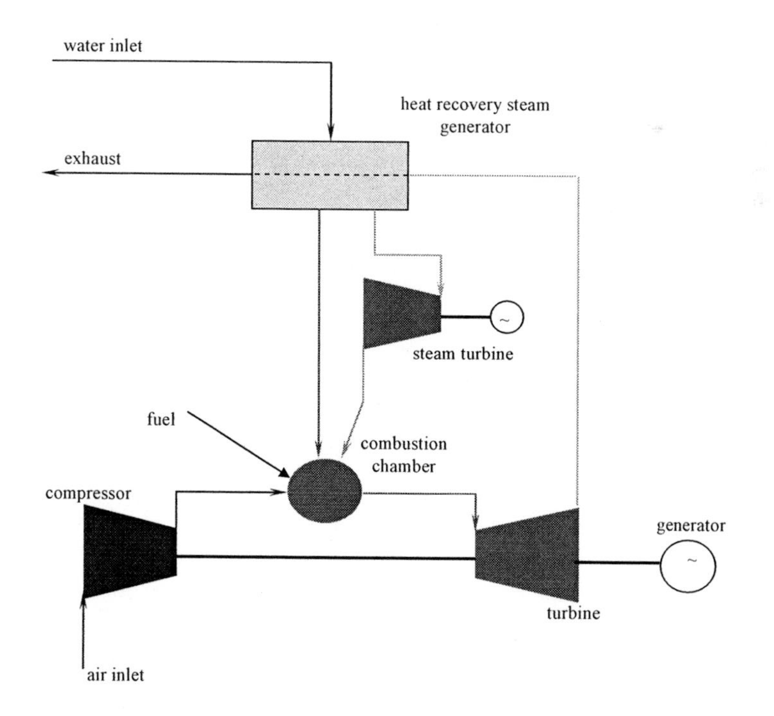

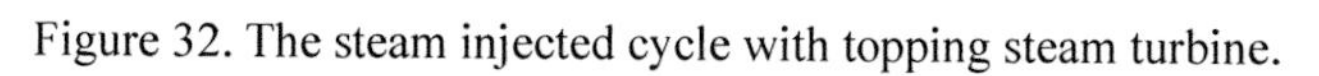
Figure 32. The steam injected cycle with topping steam turbine.

2.11.3. The Turbo Charged Steam Injected Cycle

The turbocharged steam injected gas turbine is illustrated in Figure 33. Such configuration would result in an average increase in power of 95% and an average efficiency rise from 30% to 42.6%.

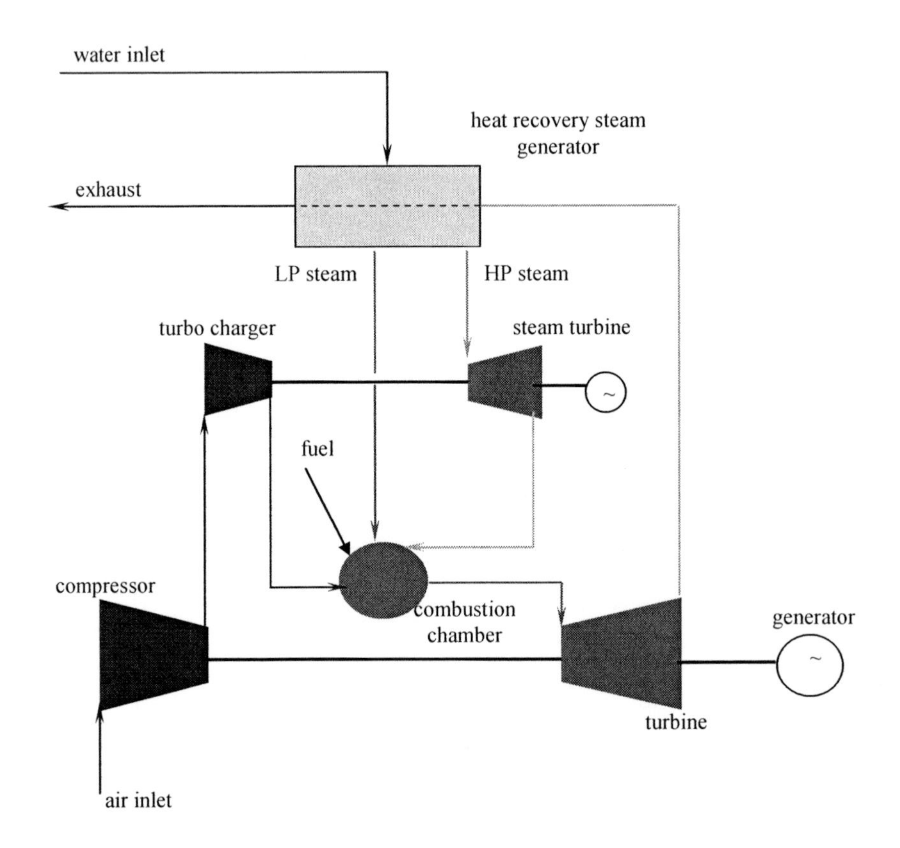

Figure 33. The turbo charged steam injected cycle.

2.11.4. The DRIASI Cycle

The Dual-Recuperated Inter-cooled – After-cooled Steam-Injected (DRIASI) cycle combines steam injection, recuperation, and water injection as shown in Figure 34. The analysis of this concept showed that the DRIASI cycle can provide comparable or superior efficiencies to those of conventional combined cycles for small systems up to 30MWe. For larger systems, the performance of the DRIASI cycle was found to be inferior both to combine cycles and to steam injected turbines.

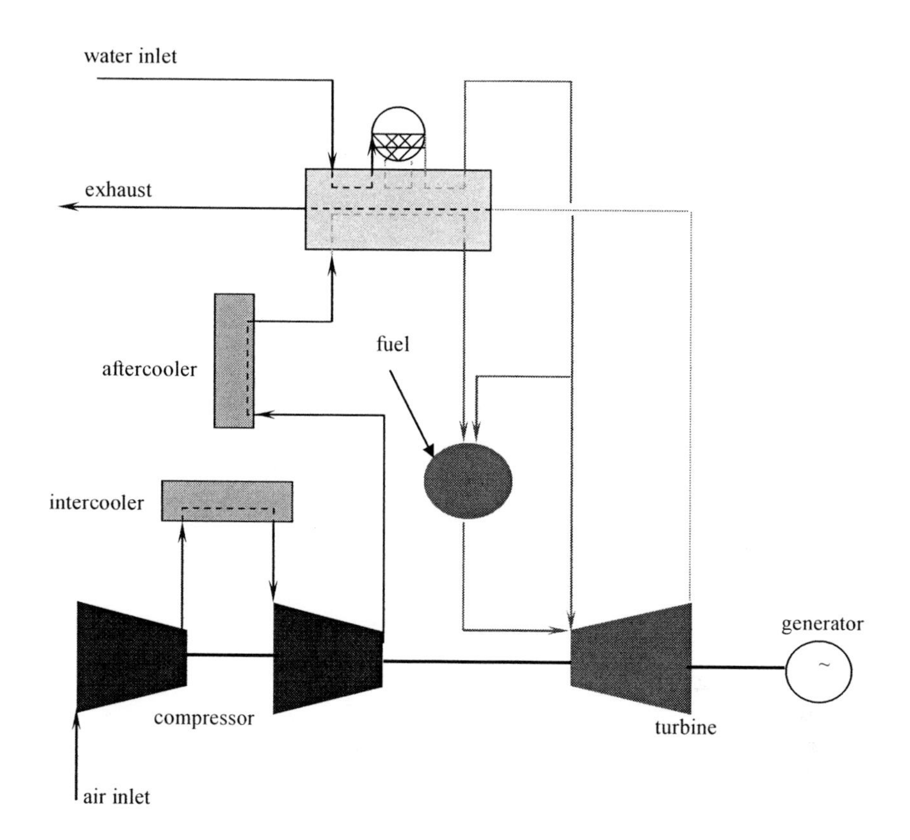

Figure 34. The DRIASI cycle.

2.11.5. The Evaporation Cycle

The limitations of the waste heat boiler can be reduced with multi-pressure systems so that the successive saturation temperatures are matched more closely to those of the exhaust gases with a better approximation to a reversible process. However, steam injection systems cannot accommodate steam flows at different pressures without substantial design convolutions. Evaporation cycles overcome the boiler limitation problem by injecting liquid water into the gas turbine air flow at the compressor exit, as illustrated in Figure 35. In effect, the heat of compression is used to evaporate the water and the resulting, single phase mixture is then heated by the turbine's exhaust gas in a suitable heat exchanger. The benefits are the same as for steam injection, i.e., higher mass flow through the turbine and increased working fluid specific heat.

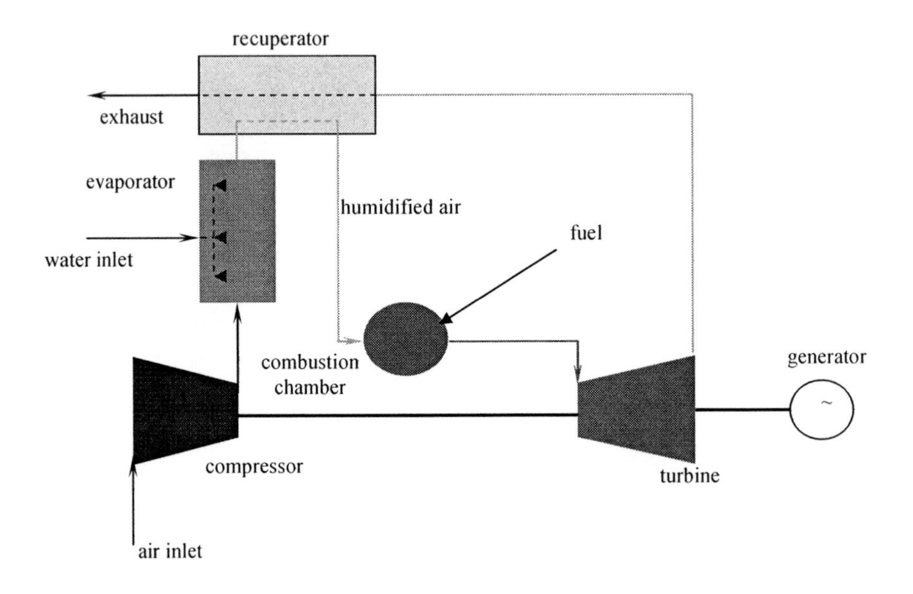

Figure 35. The evaporation cycle.

2.11.6. The HAT Cycle

The best known of the advance evaporation cycles is referred to as the humid air turbine (HAT) cycle. Originally proposed as the evaporative-regenerative cycle the HAT cycle provides a substantial power boost of the system and an efficiency rise of several percentage points. A more advanced concept with intercooling, illustrated in Figure 36, can provide even higher efficiency.

The air is first compressed in the low-pressure compressor of the gas turbine and then enters the intercooler. The heat of compression is recovered for air saturation by circulating water and make-up water, which is passed to the saturator. Two main saturator types can be used in the HAT cycle, such as, plate towers and packing towers. In the former the contact between the air and water flows is carried out by subsequent steps, because the liquid falls from one plate to the next. The latter exploits internal packing in order to enhance heat and mass exchange surface between the gas and water. The packing towers are characterized by lower pressure drops and lower capital cost, which will benefit

HAT cycle applications. The saturator can operate with any clean and filtered water source as long as the dissolved substances at the water outlet remain below their precipitation concentration at the operating conditions. The water quality is maintained via a combination of the water treatment system and the saturator blow down to purge impurity.

The cooled compressed air is further compressed in the high-pressure compressor, cooled in the aftercooler and then fed to the air saturator. The air is contacted over packing with water heated by the various heat sources. The humid air leaving the saturator, after preheating in the gas turbine exhaust, is fed to the combustion chamber. The hot gas exiting the combustion chamber expands through the turbine driving the compressors and providing power. The exhaust heat is then recovered in the recuperator and in the economizer to preheat the water for air saturation. The water content increases the mass and the specific heat of the flow, which leads to additional power, and the use of the recuperator gives higher efficiency. By varying the water content the HAT cycle can be put in part load operation without penalizing efficiency, and can also be started up in much shorter time than a conventional combined cycle plant.

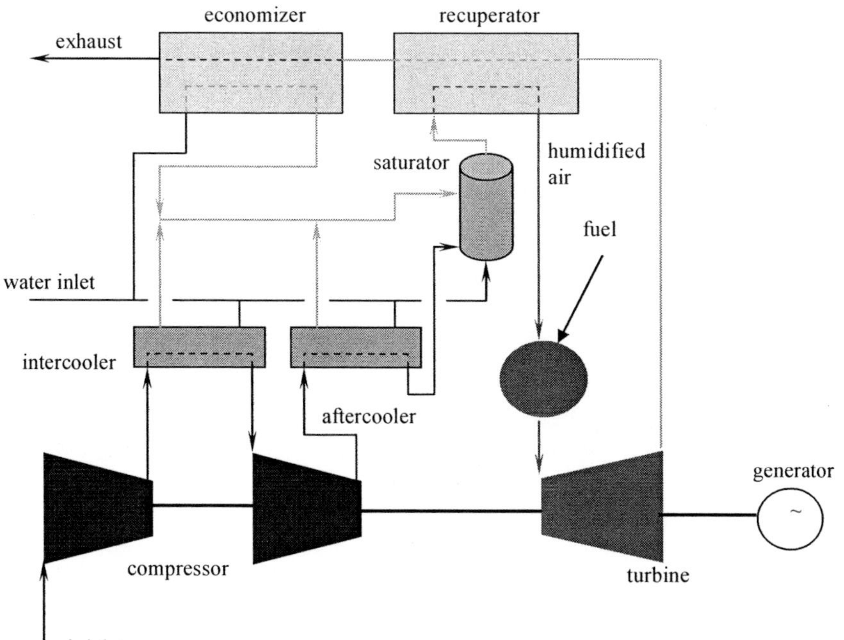

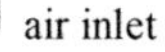

Figure 36. The HAT cycle.

Water consumption is a problem for the same reasons as apply to steam injection cycles but the consumption rate is only about one third. Based on various studies the efficiency of the HAT cycle varies from 54% for a low-pressure ratio turbine to 57% for a high-pressure cycle. Further, the cycle does not require expensive steam/water equipment that simplifies the scheme and lowers operating and maintenance costs. A HAT cycle pilot project (600kW) is currently under progress at the Lund University.

This is part of the evaporative gas turbine project, which involves Swedish Universities and main European gas turbine companies. The main aim of the project is the installation and experimentation of the first working HAT cycle in the world.

2.11.7. The LOTHECO Cycle

The LOw Temperature HEat COmbined cycle (LOTHECO) cycle, shown in Figure 37, uses a mixed air steam turbine, as a topping cycle, with an external energy source for the water-in-air evaporation (e.g., solar energy). The heat contained in the exhaust gas is utilised in a bottoming Rankine cycle. Since the water-in-air evaporation takes place at the vapour partial pressure, the saturation temperature is accordingly significantly low (below 170°C).

This temperature range (from below 100°C up to 170°C, depending on the amount of injected water and the compressor pressure ratio) is in favour of the integration of low-quality heat sources, that under other circumstances cannot be utilised for electric power generation, such as, geothermal, solar, etc. These arrangements result in an enhanced fuel-to-electricity efficiency compared to the efficiency of an equivalent conventional combined cycle. Efficiencies above 60% have been recently reported and can, therefore, be seen as a choice for the future power plants.

2.11.8. The Wet Compression Cycle

In the wet compression cycle water injection is accomplished at the compressor stages, which results in nearly isothermal compression (Figure 38). Water in the exhaust is recovered in a separating condensing unit. Such configuration can reach efficiency of 43%.

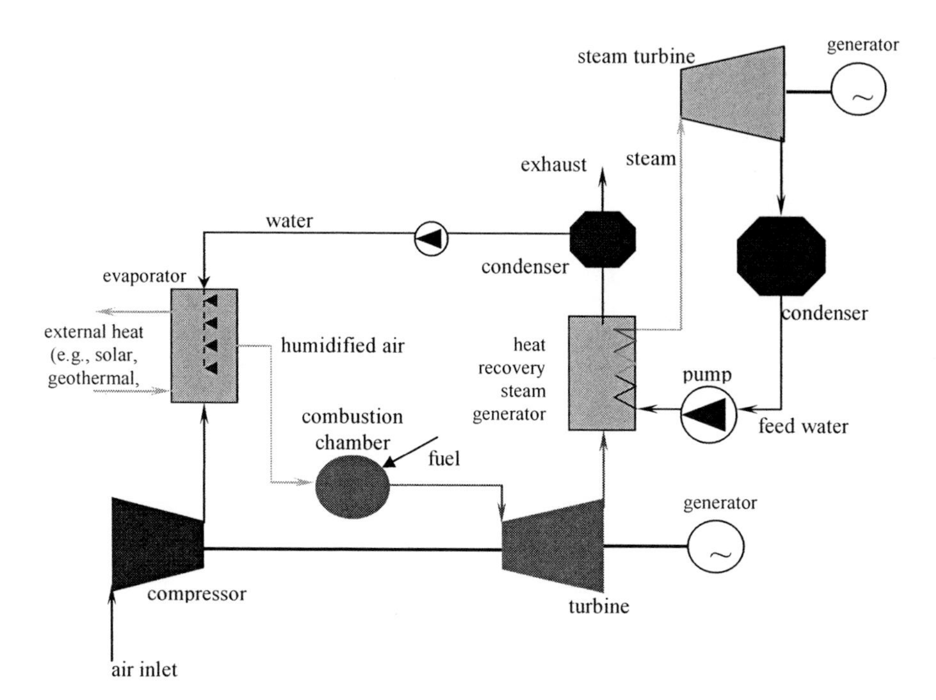

Figure 37. The LOTHECO cycle.

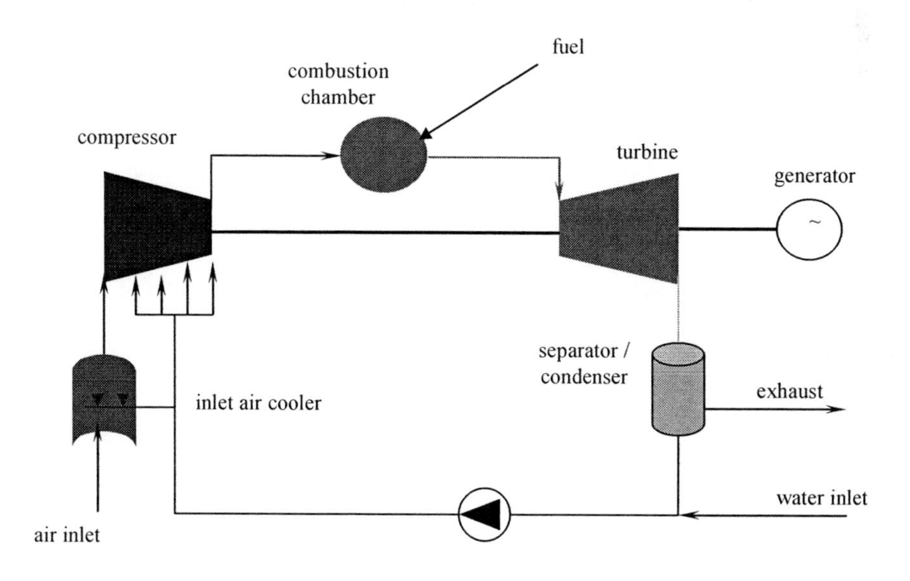

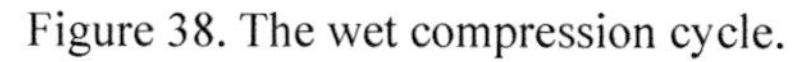
Figure 38. The wet compression cycle.

2.12. Nuclear Power

At present, nuclear reactors supply about 18% of the worldwide electric power. In these nuclear reactors, energy is generated by the fission of uranium as indicated in Figure 39(a). The nuclei of uranium are made to split into two or more pieces, a reaction that releases a large amount of energy in the form of heat. The same kind of reaction also generates the explosive energy of an atomic bomb; but whereas in an atomic bomb the energy is released in a sudden burst, in a nuclear reactor the energy is released gradually, under controlled conditions.

An alternative to fission is nuclear fusion which is an attractive source of energy because it bypasses some of the safety problems associated with nuclear fission. In fusion reactions, energy is generated by merging nuclei of hydrogen, deuterium, or tritium; these nuclei combine to form helium as presented in Figure 39(b).

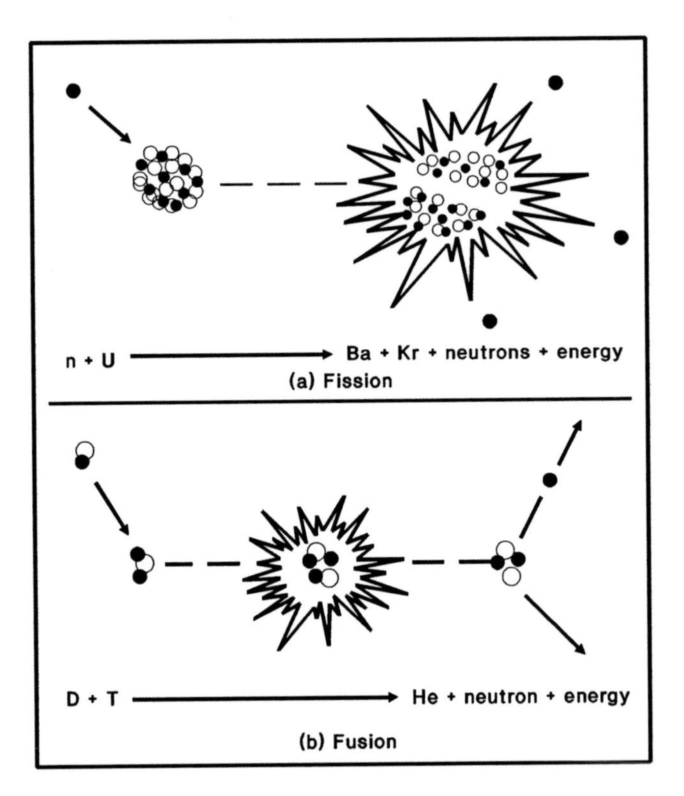

Figure 39. Fundamentals of nuclear fission and fusion reactions.

2.12.1. Nuclear Fission

Nuclear fission power plants are working on the Rankine cycle principle shown in Figure 40.

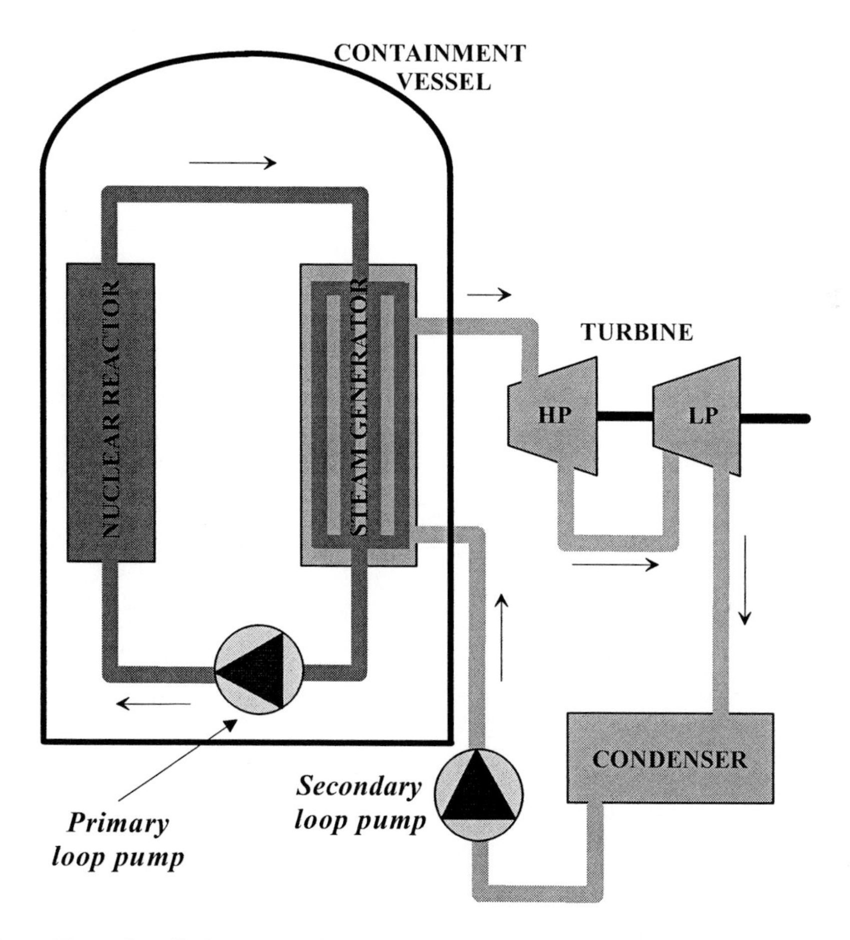

Figure 40. Nuclear fission power plant.

The total fission of 1 kg of uranium releases as much heat as the combustion of 3 million kg of coal. However, commercial reactors use only a few percent of the fresh uranium fuel placed in them since the rest must be reprocessed chemically before it can be used again. Furthermore, the efficiency of nuclear power plants is lower than that of fossil fuelled power plants since in a nuclear power plant only about 30% of the thermal energy released by the uranium is converted into electric energy, whereas in a modern fossil fuelled power plant about 40% of the thermal energy of coal or oil is converted into electric energy.

Naturally occurring uranium is a mixture of two isotopes (a) 99.3% of ^{238}U and (b) 0.7% of ^{235}U. Only the rare isotope ^{235}U will be used in an ordinary nuclear reactor. Since it is rather difficult to separate the isotopes completely, most nuclear reactors have been designed to operate with enriched uranium (several percent ^{235}U) rather than pure uranium ^{235}U. In this mixture, only the ^{235}U is used, while the ^{238}U remains behind. The worldwide supply of high-grade ores suitable for the (partial) extraction of ^{235}U is rather limited.

Fortunately, the abundant isotope ^{238}U can be converted, by nuclear alchemy, into ^{239}Pu, an artificial isotope of plutonium, which will burn in a nuclear reactor. The conversion is accomplished in a breeder reactor. In such a reactor, an initial supply of ^{239}Pu is surrounded by a blanket of ^{238}U. As the plutonium fuel burns, it releases not only heat, but it also releases an intense stream of fast neutrons which impinge on the uranium and convert it into plutonium. Thus, the reactor produces not only energy, but also new fuel, in an efficient breeder in which, the amount of new fuel generated by conversion of uranium exceeds the amount of initial fuel consumed.

Unfortunately, nuclear fission yields rather dirty energy since fission reactions generate dangerous radioactive residues. Nuclear power plants must be carefully designed to hold these radioactive residues in confinement. The cumbersome safety features that must be incorporated in the design make the construction and maintenance of nuclear power plants extremely expensive. Furthermore, when the load of fuel of a nuclear reactor has been spent, the residual radioactive wastes must be removed to a safe place to be held in storage for hundreds of years until their radioactivity has died away.

2.12.2. Nuclear Fusion

Fusion —the energy source of the sun— has the long-range potential to serve as an abundant and clean source of energy. The basic fuels, deuterium (a heavy form of hydrogen) and lithium, are abundantly available to all nations for thousands of years. There are no emissions from fusion, and the radioactive wastes from fusion are short lived, only requiring burial and oversight for about 100 years. In addition, there is no risk of a melt-down accident because only a small amount of fuel is present in the system at any time. Finally, there is little risk of nuclear proliferation because special nuclear materials, such as uranium and plutonium, are not required for fusion energy.

Fusion systems could power an energy supply chain based on hydrogen and fuel cells, as well as provide electricity directly. Although still in its early stages of development, fusion research has made some advances. Internationally, an effort is underway in Europe, Japan, and Russia to develop plans for constructing a large-scale fusion science and engineering test facility. This test facility may someday be capable of steady operation with fusion power in the range of hundreds of megawatts.

Note that the process of fusion is the reverse of fission, that is, light nuclei (such as hydrogen) release energy when they merge, whereas in fission heavy nuclei (such as uranium) release energy when they split.

In order to initiate a fusion reaction, the nuclei must be subjected to extremely high temperatures and pressures. For instance, the fusion reactions at the center of the Sun involve a temperature of 15 million K and a pressure of 1 billion bar. Nuclear reactions that require such extreme temperatures are called thermonuclear. Schemes for the development of fusion power on Earth intend to use deuterium or tritium because their reaction rates are much faster than that of hydrogen. Under these conditions, the deuterium or tritium will be in the form of plasma and it cannot be contained by a conventional reactor vessel of steel or ceramic material. One scheme for fusion attempts to suspend the plasma in the middle of an evacuated vessel by means of magnetic fields; the development of this scheme is the objective of most of the modern plasma research as indicted in Figure 41.

2.13. THE INTERNAL COMBUSTION ENGINE

Internal combustion engines are available from small sizes (e.g., 5kWe for residential back-up generation) to large generators (e.g., 25MWe) and they commonly use available fuels such as natural gas and gasoil.

An internal combustion engine converts the energy contained in a fuel into mechanical power. This mechanical power is used to turn a shaft in the engine. A generator is attached to the internal combustion engine to convert the rotational motion into power. There are two methods for igniting the fuel in an internal combustion engine. In spark ignition, a spark is introduced into the cylinder (from a spark plug) at the end of the compression stroke. Fast-burning fuels, like natural gas, are commonly used in spark ignition engines. In compression ignition, the fuel-air mixture spontaneously ignites when the compression raises it to a high-enough temperature. Compression ignition works best with slow-burning fuels, like gasoil.

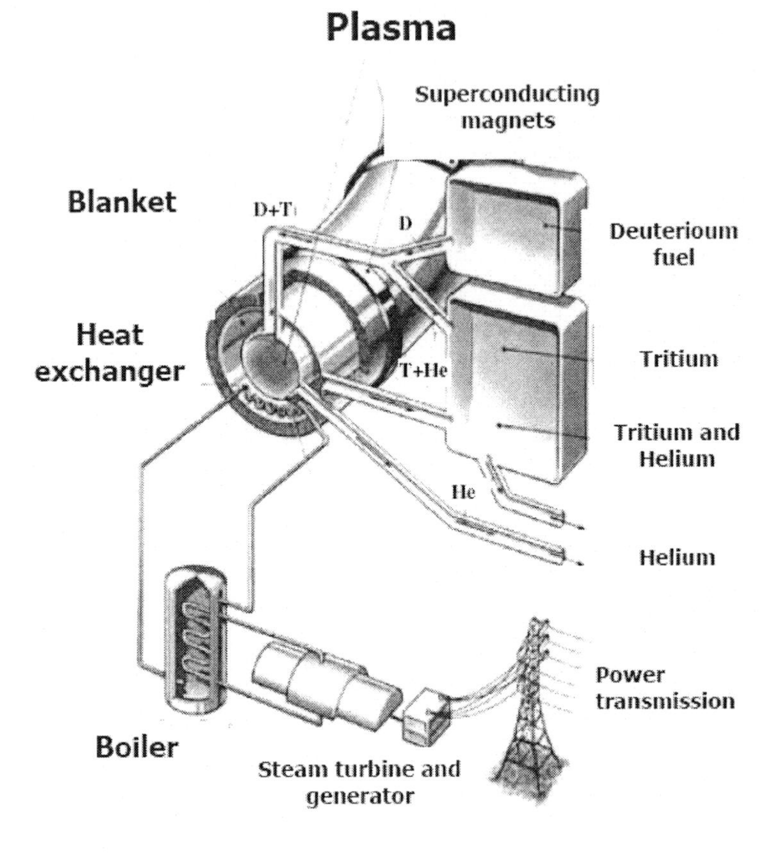

Figure 41. A future nuclear fusion power plant.

An internal combustion engine is operated in two main cycles. The four stroke cycle and the two stroke cycle. In the four stroke cycle each movement of the piston up or down the cylinder is a stroke. The four stroke cycle consists of an induction stroke where air and fuel are taken into the cylinder as the piston moves downwards, a compression stroke where the air and fuel are compressed by the upstroke of the cylinder, the ignition or power stroke where the compressed mixture is ignited and the expansion forces the cylinder downwards, and an exhaust stroke where the waste gases are forced out of the cylinder. The intake and outlet ports open and close to allow air to be drawn into the cylinder and exhaust gases to be expelled. In the two stroke cycle the crankshaft starts driving the piston toward the spark plug for the compression stroke. While the air-fuel mixture in the cylinder is compressed, a vacuum is created in the crankcase. The crankcase is creating a vacuum to suck in air/fuel from the carburetor through the reed valve and then pressurizing the crankcase so that air/fuel is forced into the combustion chamber. This vacuum opens the reed valve and sucks air and fuel from the carburetor. Once the piston leads to the end of the compression stroke, the spark plug fires to generate combustion pressure to drive the piston. The sides of the piston are acting like valves, covering and uncovering the intake and exhaust ports communicating into the side of the cylinder wall. Two stroke engines are lighter, simpler and less expensive to manufacture. They have a greater power to weight ratio, but they are lesser in efficiency and they require lubrication oil to be fed with fuel.

Chapter 3

3. CARBON CAPTURE AND STORAGE TECHNOLOGIES

Current scenarios and projections to 2030 indicate that there will be an increase in worldwide energy demand; therefore, there is clear evidence that in the short and medium term horizon fossil fuel resources, such as, coal, oil and gas, will continue to dominate. During the above transition to a long term fully sustainable energy economy, hydrogen is likely to be produced mainly from fossil fuels. Consequently, unless specific policy initiatives and measures are undertaken, global greenhouse gas emissions will rise by an unacceptable 60% before 2030. It has been estimated that without any action taken to reduce greenhouse gas emissions, the global average surface temperature is likely to rise by 1.8°C - 4.0°C by the end of this century. Even the lower end of this range would take the earth's temperature above the threshold value beyond which irreversible and possibly catastrophic changes become far more likely. Projected global warming this century is likely to trigger serious consequences for humanity and other life forms, including a rise in sea levels of between 18cm and 59cm, which will endanger coastal areas and small islands, and a greater frequency and severity of extreme weather conditions.

CO_2 is considered to be the principal greenhouse gas responsible for global warming. The challenge presented is the reduction of CO_2 emissions by 50% - 80% between now and 2050. It is widely acknowledged that CO_2 capture and storage (CCS) technologies can play an important role in mitigating CO_2 emissions. Facing this life threatening phenomenon, it is essential that CCS technologies should receive the appropriate attention required in order to constitute a promising option for gradual de-escalation of this serious problem.

It is clear that the commercialisation of zero emission power plants is not viable in the short term and so a range of technologies and options for different fuels should be explored, rather than adopting a single approach to achieve zero emissions. In addition, the development of cross-cutting technologies such as new high temperature and corrosion resistant materials, advanced control systems and more efficient cooling schemes will be beneficial to all the future zero emission plants, making them more attractive to utility companies, particularly in the first stages of the learning process. Each technology option that shows significant promise will have to proceed to large-scale demonstration at the earliest opportunity.

The main technologies used to generate power from fossil fuels are pulverised coal cycles and natural gas combined cycles, while IGCC technology is an emerging technology indicating encouraging results on CO_2 capture. The available CCS technologies, shown in Figure 42, could be incorporated in all of these types of plants, either in oxyfuel combustion or in a pre- or post-combustion process. The capture of CO_2 in power plants will come at a price, since it will require energy from the plant. This will not only affect the cost of electricity, it will also increase the use of fossil fuel at a plant level.

In a pre-combustion CCS process the fossil fuel is partially oxidized in a reactor with steam and air. In a second shift reactor, the resulting CO reacts with steam to produce additional H_2. The result is a mixture of H_2 and CO_2. The CO2 stream is sent to compression and dehydration and the H_2 can be used as fuel.

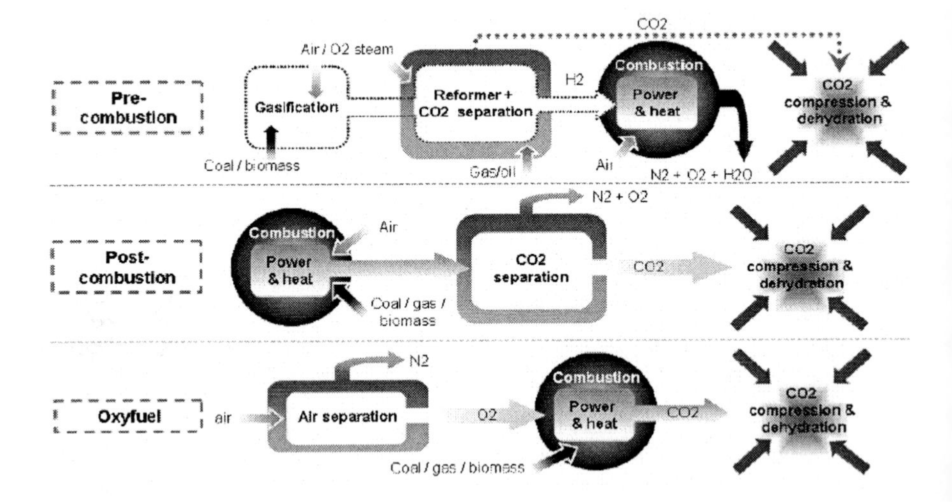

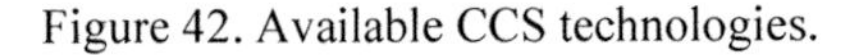
Figure 42. Available CCS technologies.

In a post-combustion CCS process CO_2 is removed after combustion of the fossil fuel. CO_2 is captured from N_2 rich flue gases using an organic/inorganic solvent to separate CO_2. N_2 and O_2 escape as exhaust. The CO_2 is then sent to compression and dehydration. In an oxyfuel combustion CCS process the primary fuel is burned in O_2 instead of air. The flue gas consists mainly of CO_2 and water vapour, the latter of which is condensed through cooling. The result is an almost pure CO_2 stream that can be transported to the sequestration site and stored.

Saline aquifers have by far the greatest potential for storing CO_2. Such aquifers are sedimentary rocks (usually sandstone and less frequently limestone or other rocks), which are porous enough to store great volumes of CO_2 and permeable enough to allow the flow of fluids. Storage of CO_2 will take place at depths below some 7m-800m where CO_2 behaves as a fluid, and where the pores of the sediments are filled with salt water. Also, depleted oil and gas fields present a significant possibility for CO_2 storage.

An attractive storage method is to inject CO_2 into mature oil fields to improve the recovery of oil (and gas) through Enhanced Oil Recovery (EOR), increasing production by 4%-20%. Finally, un-mineable coal seams offer another opportunity to store CO_2 at a low net cost. While this approach is still in its early stages and needs more research, it is considered a promising concept due to the added value of the produced methane.

In this chapter, a brief overview of the most common power generation technologies which can incorporate CCS systems, such as pulverised coal technologies and natural gas combined cycle technologies are briefly described. Also, the IGCC and oxyfuel technologies which appears to have great potential with the integration of a CCS system are presented.

3.1. The Pulverized Coal Technology with CCS

The pulverized coal plants can employ a post-combustion CCS process in order to minimize harmful CO_2 emissions to the environment. The process is based on a number of flue gas CO_2 amine scrubbers. These scrubbers employ monoethanolamine as a scrubbing agent to absorb CO_2 and then release it in a steam-heated generator. The released CO_2 is routed to the CO_2 compressor station for export. The operating experience of existing plants has shown fairly high operating costs and efficiency penalization due to high steam consumption and monoethanolamine makeup. Net plant output also decreases as power is internally supplied to the CO_2 scrubber plant and compressor stages. A typical pulverized coal plant with a post-combustion CCS process is shown in Figure 43.

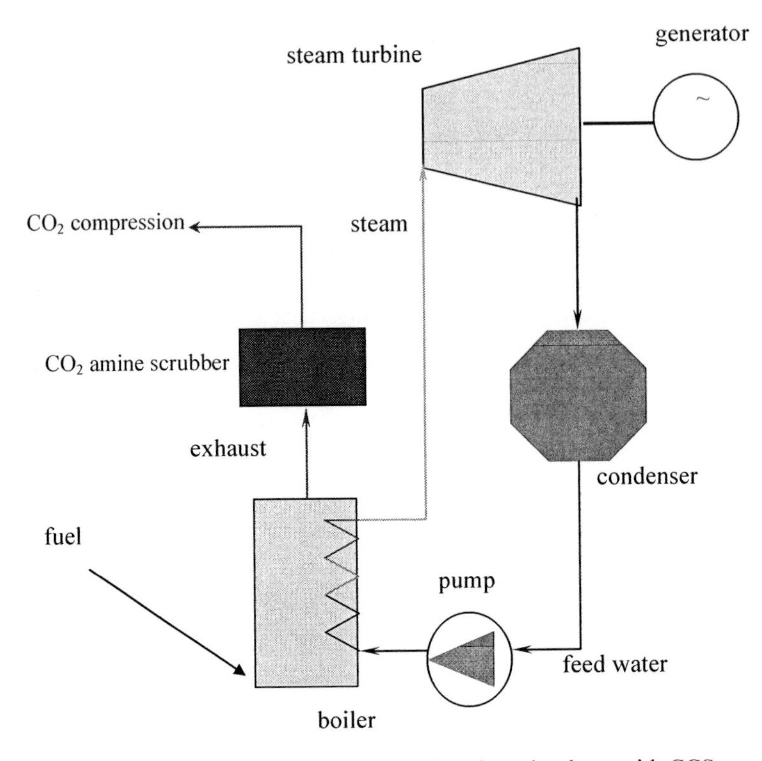

Figure 43. General arrangement of pulverised coal cycle technology with CCS.

3.2. THE IGCC CYCLE

It is internationally acclaimed today that the technology of Integrated Gasification Combined Cycle (IGCC) is one of the most important bulwarks of the future hydrogen economy which HYPOGEN plants will be based on. The scientific community and the major corporations of electricity generation consider the technology used in IGCC as one of the most promising technologies for the production of cleaner electrical power in the future. This is underlined by the increasing amount of funding available for research projects that are closely related to electricity generation with the use of hydrogen (or synthesis gas) and by the development of joint ventures between major corporations in the energy sector to fund the construction and operation of IGCC-based hydrogen power plants across Europe.

It should be noted that although each major component of IGCC has been broadly utilized in industrial and power generation applications, the integration of a gasification island with a combined cycle power plant is relatively new. This integration for commercial electricity generation has been demonstrated by a number of facilities around the world, but is not yet perceived to be a mature, commercial technology with clearly understood costs and risks.

IGCC technology is a power generation process that integrates a gasification system with a combined cycle power plant. The gasification system converts coal (or other carbon-based feedstocks such as petroleum coke, heavy oils, biomass, etc.) into synthesis gas (syngas) which consists primarily of hydrogen (H_2) and carbon monoxide (CO). The syngas (H_2/CO) is then used as fuel in a combined cycle power plant for electricity generation.

A typical configuration of an IGCC power plant without carbon capture and storage (CCS) is illustrated in Figure 44 and with CCS is presented in Figure 45. It comprises mainly of four operating components, such as, the air separation unit, the gasifier, the syngas cooling and clean-up system and the combined cycle power plant.

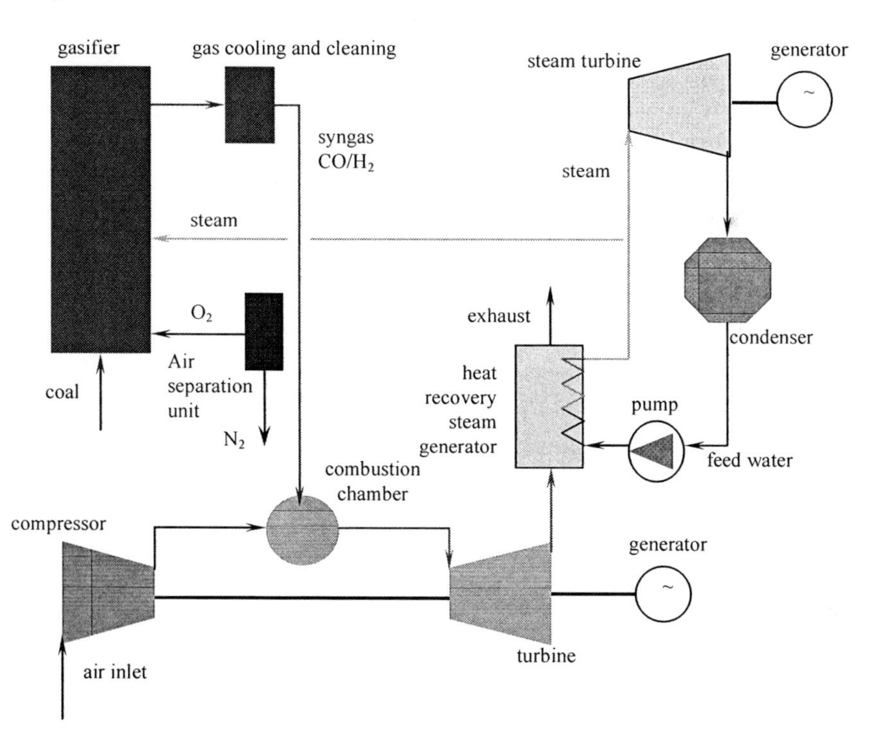

Figure 44. General arrangement of IGCC technology without CCS.

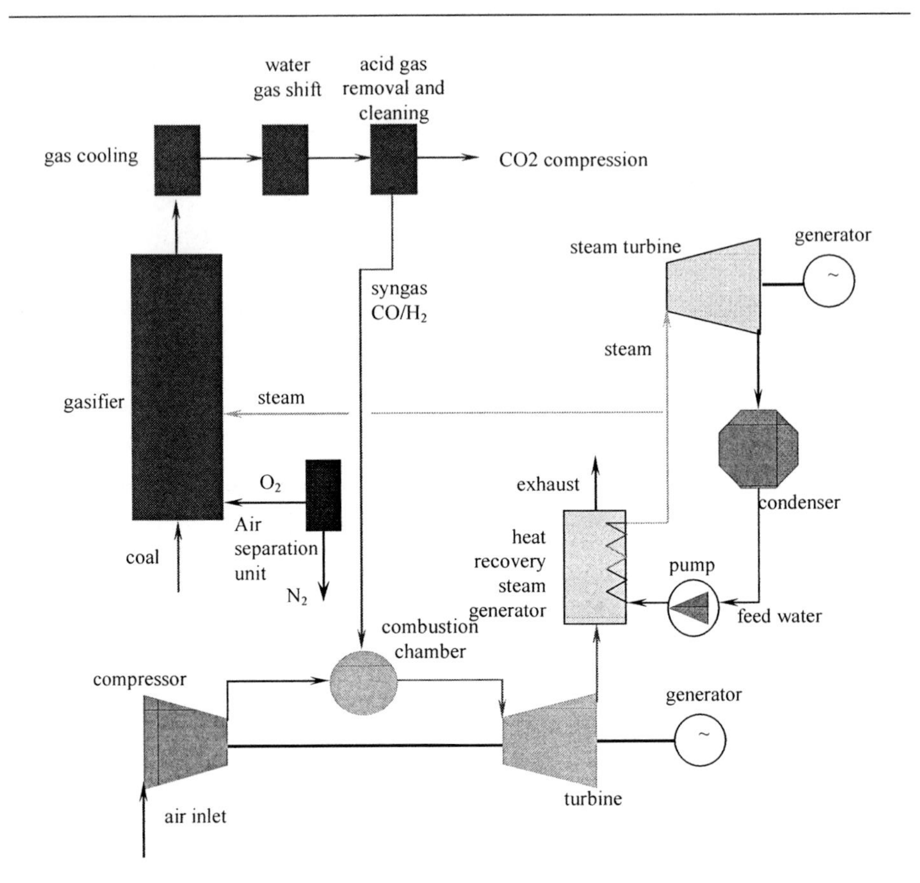

Figure 45. General arrangement of IGCC technology with CCS.

The pressurized cryogenic air separation unit is responsible for the separation of air into its constituents and the supply of pure oxygen to the gasifier. The pressurized air required by the air separation unit can be supplied either entirely by the gas turbine compressor or entirely by a separate compressor or partially from the two compressors. The degree of integration between the air separation unit and the gas turbine is defined as the portion of the air required by the air separation unit that is extracted and supplied by the compressor of the gas turbine. In the case of full (100%) or partial integration, the nitrogen produced from the air separation unit is used as the syngas diluent prior to combustion. Thus, nitrogen is injected into the gas turbine's combustor burner and mixed with the syngas in order to control NO_X emissions, reduce the tendency for flashback during combustion and enhance the power output from the turbine.

In the case of zero degree of integration, the preferred diluent is either steam or water since the nitrogen originating from the air separation unit would be at low pressure and would require additional compression prior to mixing it with syngas. The degree of integration can play a significant role in the overall performance and efficiency of the power plant. Clearly, as the degree of integration increases and therefore the air separation unit is supplied with air from the gas turbine compressor, the net plant output power increases since the auxiliary power requirements of the air separation unit become less. The gas turbine can still maintain its rated power output with the reduced amounts of supplied gases. This is because syngas, due to its necessary dilution and chemical properties, has a much higher volumetric and mass flow rate in the combustor than natural gas. Therefore, contemporary gas turbines that are built for use with natural gas would require fewer amounts of combusted syngas to keep their rated power output. At some stage however, and as the degree of integration increases, the amount of air diverted towards the air separation unit starts to be sufficiently high to negatively affect gas turbine output performance. Thus the net plant output power starts to be reduced.

Clearly, the degree of integration where the maximum net power is reached is not the same to where the maximum plant efficiency is reached. Maximum plant efficiency is reached at 100% integration, where the auxiliary power requirements of the air separation unit are minimal since all air is supplied from the gas turbine compressor. The degree of integration where the maximum plant output power is reached varies according to the type of the gas turbine used and the type of coal and gasifier used.

Oxygen-blown gasifiers, at high temperature and elevated pressure in the presence of oxygen and steam, convert carbon-based feedstocks into syngas (H_2/CO). Gasification is a partial oxidation process that occurs in reducing conditions (controlled shortage of oxygen) inside an enclosed pressurized reactor. Partial oxidation of the feedstock creates heat and a series of chemical reactions produce syngas. The alternative to oxygen-blown gasification is the air-blown gasification which eliminates the need for an air separation unit. However, air-blown gasification has the disadvantage of producing a syngas with lower calorific value, which is mostly not desirable and thus is not preferred.

The gasifiers can be divided into three groups namely, (a) the moving-bed reactors (also called fixed-bed), (b) the fluidized bed reactors and (c) the entrained flow reactors. In moving-bed reactors large particles of the fuel move slowly down through the gasifier while reacting with the gasifying medium moving up through it. Several different reaction zones are created that accomplish the gasification process. Operating temperatures are not uniform inside the reactor with the temperature of the syngas leaving the reactor being as low as 400-500°C. In fluidized-bed reactors small particles of the fuel remain suspended in the gasifying medium while the gasification process takes place. The temperature inside the reactor remains uniform in the range of 800-1000°C. In entrained flow reactors the pulverized fuel goes through the various stages of gasification flowing co-currently with the gasifying medium. The feedstock can be either in dry or in water slurry form. The temperatures achieved in the reactor are very high in the range of 1200-1600°C. Entrained-flow gasifiers are considered to be the most suited type for IGCC applications.

The raw syngas produced from the gasifiers contain (apart from CO and H_2) small quantities of carbon dioxide (CO_2), water (H_2O), nitrogen (N_2), sulfur compounds (H_2S, COS), nitrogen compounds (NH_3, HCN), chlorine compounds (HCl, NH_4Cl), particulate matter (unconverted carbon and ash), methane (CH_4), mercury (Hg), etc. Prior to injecting the syngas in the combustor of the gas turbine the raw syngas has to be treated in order to be cleaned up. Due to the fact that the syngas leaves the gasifier at high temperatures it has to be cooled down first.

Cooling of syngas is accomplished using a waste heat boiler, or a direct quench process that injects water into the raw syngas. For particulate removal, ceramic or metallic filters are used located upstream of the heat recovery device, or by water scrubbers which are located downstream of the cooling devices. Next, the syngas is treated in "cold-gas" clean up process (called acid gas removal process) to remove the sulfur and nitrogen compounds. This process can be either chemical or physical solvent-based process. Other compounds such us HCl and NH_3 are removed using water scrubbing. Traces of mercury are also removed from syngas using a sulfur-impregnated activated carbon bed.

After clean up, the syngas is directed to the combined cycle power plant. The syngas is used as fuel in the combustor of the gas turbine. The hot exhaust gases produced after combustion of syngas are captured and directed to a heat recovery steam generator to generate steam which, in turn, is fed to a steam turbine to complete the combined power cycle. The exhaust gases leave the heat recovery steam generator through a stack in the atmosphere. Due to the clean-up of the syngas prior to combustion these exhaust gases are environmentally friendly containing much smaller amounts of SO_2, NO_X, CO_2 and particulate matter compared with other coal-based technologies.

It is acknowledged that greenhouse gas emissions from an IGCC plant are much lower than those from a typical pulverized coal plant. However, the need for further reduction of harmful emissions to the environment (Kyoto protocol) and the developments in technology have advanced the idea of an IGCC plant incorporating a CCS cycle. Such a plant would provide minimal harmful emissions while still having the advantages of a standard IGCC plant (cheaper and abundant fuel, marketable by-products) plus the possibility of pure hydrogen co-production for storage or fuel cell applications. Essentially such a plant would embody the HYPOGEN principle as envisaged by the EU.

Recent studies have shown that an IGCC plant with CCS requires two additional pre-combustion stages than the conventional IGCC cycle plant as illustrated in Figure 45. The two additional stages are the water gas shift reaction and the acid gas removal for the removal of CO_2 from the syngas. In addition, a CO_2 compression stage is necessary to make transportation and storage of the sequestered quantity of CO_2 feasible. The downside is that these additional stages reduce overall plant efficiency when compared to the IGCC plant without CCS. Plant efficiency partly decreases because internal power is required to drive the CO_2 compression stage. Another reason for efficiency decrease is that after the superimposition of the two additional stages (before syngas combustion), the amount of coal feed required to provide the necessary rate of chemical fuel energy to the gas turbine needs to be increased. In turn, this can result in lower steam/carbon ratio in the gasifier which would necessitate the supply of additional steam from the stream cycle and thus lower plant output power even further. The amount of efficiency penalty for the IGCC plant with CCS also depends heavily on the type of gasifier used.

Another important factor is the carbon capture efficiency of the plant. Clearly, the degree of capture efficiency (which refers to the percentage amount of CO_2 to be captured from the plant) will influence the final cost of electricity and therefore needs to be decided in advance. Although no commercial IGCC plant with CCS has been built yet, the typical value of capture efficiency considered in recent studies is 90%. One factor that may limit the level of capture efficiency of the plant is that contemporary gas turbines are not guaranteed to function with fuel syngas having more than 65% hydrogen content. Clearly with higher capture efficiencies, the hydrogen content of syngas increases to levels that cannot be tolerated for electricity production.

The water gas shift process is described briefly below. During the water gas shift process the exothermic water gas shift reaction transfers the fuel heating value from CO to H_2 and transfers the carbon from CO to CO_2. This changes the chemical composition of the syngas towards more H_2 and less CO. There are two important reasons that justify the use of this additional stage in the process of CO_2 capture. The first reason is that the conversion of CO into H_2 increases the total amount of hydrogen production for a given quantity of input fossil fuel. The second reason is that the available quantity of CO is minimized since it is effectively being converted into CO_2. By this the capture efficiency of CO_2 is increased and CO_2 can be separated from hydrogen much more easily than CO, therefore this enhances the process of CO_2 sequestration and pure hydrogen production. For maximum conversion of CO and therefore maximum hydrogen production, low temperatures are applied in the shift reactors. The raw syngas is therefore cooled after it exits the gasifier so that it reaches typical temperatures between 200^0C and 500^0C. It then reacts with water vapor or steam in the shift reactor in the presence of H_2S. For maximum CO conversion, the shift reaction is repeated in a second stage/reactor which is at a lower temperature than the first. The syngas exiting the water gas shift reaction has a much higher composition of hydrogen than the raw syngas but still requires to be treated for CO_2 and sulfur (H_2S) sequestration. The water gas shift stage can be employed in the sour shift conversion or in the clean shift conversion mode. In the sour shift, which is the preferred mode of operation, the shift reaction stage takes place before the syngas acid gas removal stage (Figure 45). Experimental studies have shown that the sour shift mode of operation is more economical and efficient than the clean shift mode whereby the water gas shift stage is employed after the syngas sulfur removal stage.

3.3. The Natural Gas Combined Cycle Technology with CCS

A natural gas combined cycle may operate both in post-combustion and in pre-combustion CCS mode. In post-combustion mode, an amine-based scrubber system is employed as in the case of the pulverized coal plant to absorb CO_2 from the exhaust. A typical natural gas combined cycle with a post-combustion CCS process is shown in Figure 46.

The pre-combustion capture is a much more complex and expensive procedure and is the object of intensive research and technological development efforts in an attempt to reduce operating costs in the near future. Figure 47 illustrates the major processes that form part of a natural gas combined cycle with pre-combustion CCS.

These processes are explained below. For the pre-combustion capture, a natural gas reforming block is central to the whole process. Natural gas reforming is different to the coal gasification process described in the IGCC cycle. Essentially the natural gas reforming is an endothermic reaction and therefore requires an external heating source in contrast to gasification which is an exothermic reaction.

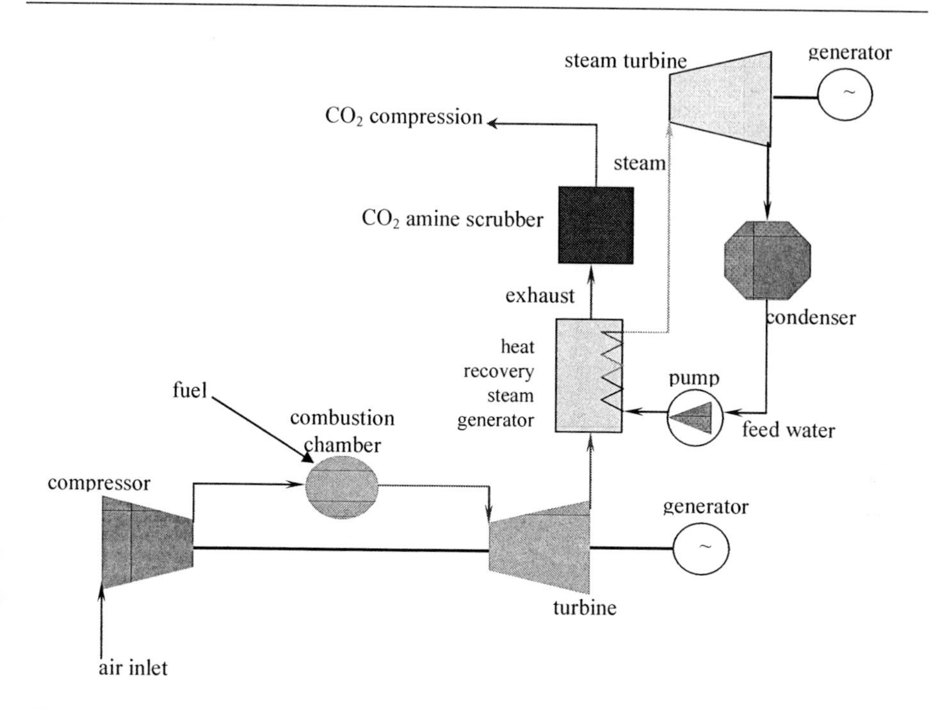

Figure 46. General arrangement of natural gas combined cycle technology with post combustion CCS.

Usually high temperature steam from the heat recovery stem generator is employed as such a source in the reformer. The operating temperature and pressure in the reformer is much less than the respective ones developed in the gasifier. Typically the reformer operates with high temperatures between 700°C to 1000°C and pressure up to 30 bar during steam reforming and with temperatures higher than 1000°C and possibly higher pressures during autothermal reforming.

The lower operating temperatures however, mean that the amount of CO converted to H_2 is higher in the reforming process which leads to the production of syngas with higher H_2 concentration than in gasification.

Also the presence of nickel-based catalyst in the reforming process limits the range of potential feedstock to only the lighter hydrocarbon fossil fuels in contrast to gasification which can have heavier hydrocarbons as feedstock. In a natural gas combined cycle with pre-combustion CCS, the natural gas can be reformed in two different ways, such as, steam reforming or authothermal reforming. Autothermal reforming is a combination of the classic steam reforming with partial oxidation (which is used in the gasification process).

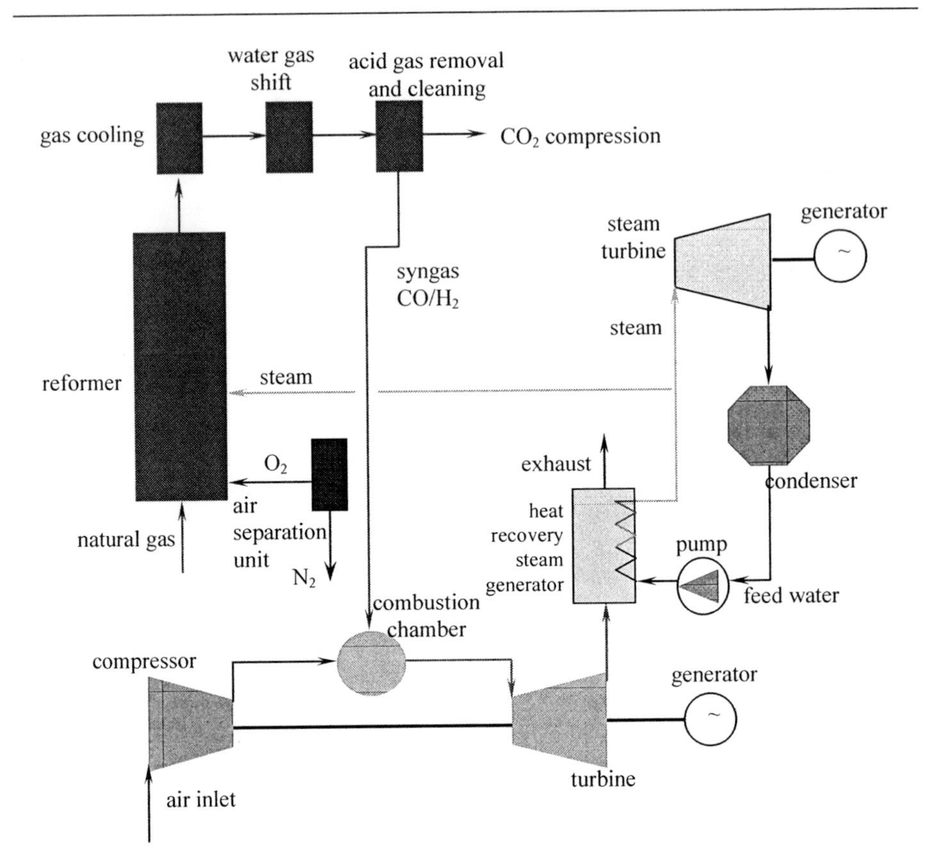

Figure 47. General arrangement of natural gas combined cycle with pre-combustion CCS.

Autothermal reforming requires less external heat source input than steam reforming and therefore is a less complex process that may have lower efficiency penalization in the steam cycle of the CC power block. Autothermal reforming is usually employed in cases where a high syngas plant output is required, for example in electricity generation purposes, as it is more economical than the steam reforming process. However, authothermal reforming typically produces lower heating value syngas than steam reforming.

The syngas from the reforming process is then treated in a two stage water shift reactor process to convert CO to CO_2 and then to a chemical absorption process to absorb and separate the CO_2. The methods used in these processes are the same as already described in the IGGC plant with CCS. After the CO_2 capture and sequestration, the syngas is used as fuel in the combustor chamber of a gas turbine in a combined cycle power plant.

3.4. OXYFUEL COMBUSTION

Oxyfuel or O_2/CO_2 recycle combustion is a relatively new technology for CO_2 capture and storage (CCS) and is still at an early development stage compared to either the post- or the pre- combustion CCS technologies. However, oxyfuel technology promises to be an economical alternative coupled with the possibility of higher CO_2 capture rates.

The principle of oxyfuel combustion can be seen in Figure 48. Basically, oxyfuel combustion is the combustion of pulverized coal or other hydrocarbon fuels, in nearly pure oxygen environment instead of air which is the conventional method employed in coal-fired steam power plants. Specifically, oxygen (of 95% purity or higher) is fed to the boiler via a cryogenic air separation unit (ASU). In addition to oxygen, the major part of the CO_2-rich exhaust flue gas is also recycled back to the boiler as a form of diluent, in order to control combustion temperature and reduce NO_x formation. This is necessary, since combustion of coal in pure oxygen gives a high flame temperature which will enhance the formation of NO_x in the boiler. Pulverized coal or other hydrocarbon fuel is then combusted in this mixture of O_2 and CO_2 within the boiler.

The major advantage of burning coal in such a mixture instead of air is that it produces a CO_2-rich flue gas that is almost nitrogen free. This flue gas composition translates to a simpler CO_2 capture and sequestration process compared to the post- or the pre- combustion CCS technologies.

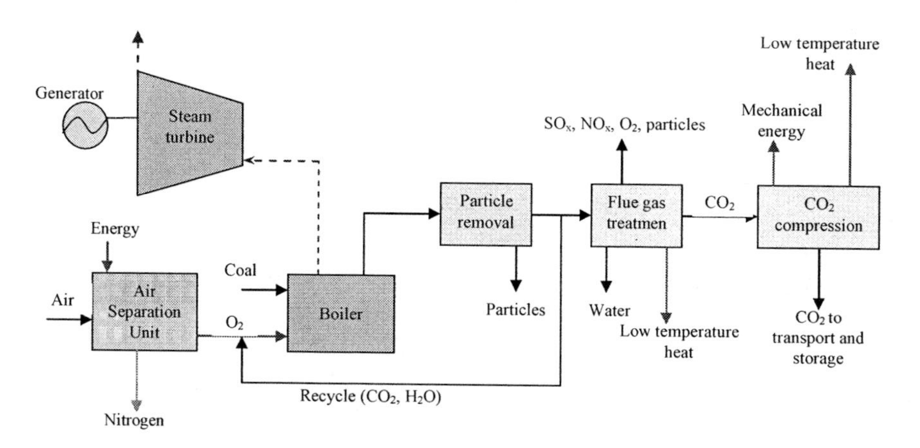

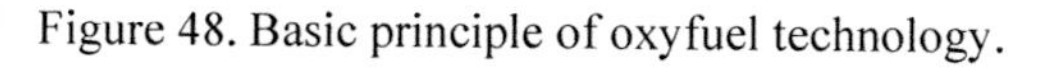

Figure 48. Basic principle of oxyfuel technology.

This advantage is amplified by the fact that the other major component of the flue gas in oxyfuel combustion is condensable water vapour. These factors can provide the foundation for an almost zero-CO_2 emission power plant (capture rate of 95% or more), something that cannot at this moment be achieved in the cases of the other CCS technologies already investigated. Another advantage arising from the flue gas composition, is that the reduced presence of nitrogen in the flue gas means that the formation of harmful NO_x emissions is significantly reduced and that plant equipment used for the flue-gas desulphurization (FGD) and nitrogen oxide removal (deNO_x) will be smaller in volume, cheaper and less complex than corresponding equipment in a conventional coal fired power plant.

Oxyfuel combustion technology produces a flue gas that is essentially composed mostly of CO_2. Final flue gas composition at the exit of the boiler may vary depending on the type of combusted hydrocarbon fuel used (e.g., pulverized coal, dry or raw lignite etc). However, typical flue gas composition can be the following: 55-65% CO_2, 25-35% H_2O, the rest being nitrogen, oxygen and argon. This composition can change depending also on the moisture levels and the nitrogen composition of the fuel used, however CO_2 remains by far the principal gas. Coal combustion can produce flue gas of lower water composition compared to lignite. In all cases, CO_2 sequestration and capture is achieved by the cleaning of the oxyfuel combustion flue gas using a series of cleaning processes. The final product gas is reportedly composed by as much as 96% CO_2, which can then be compressed and transported for storage or Enhanced Oil Recovery (EOR) applications. The basic cleaning processes are the cooling and condensing of the water vapour and the desulphurization. However, in most cases the use of gas driers for excess moisture dehydration and the use of an inerts removal stage for the removal of the non-condensable gases such as nitrogen, oxygen and argon are also employed.

The first cleaning process of the flue gas after exiting the boiler is the particle removal process via the use of dedicated filters which are usually electrostatic or fabric. This process removes the ash present in the flue gas and it is only after this stage that a part of the flue gas can be recycled back to the boiler. The required amount of flue gas to be recycled back to the boiler can only be determined empirically. The critical factor that limits the amount of recycled flue gas is flame/burnout stability within the boiler. There are two major reasons for which reduced amounts of flue gas recycle should take place in the oxyfuel process. First, excess amounts of recycled CO_2 can negatively affect flame and burnout stability by decreasing flame speed and temperature. Essentially a significant increase of oxygen concentration is required in order to compensate for the higher heat capacity of CO_2 thus keeping stable flame temperature and providing a stable combustion process. Second, the amount of recycling taking place in the process is inversely proportional to the oxyfuel plant efficiency. Therefore, by keeping the amount of recycled flue gas as low as possible, plant efficiency receives minimum penalization. The ideal level of recycled flue gas to the boiler is still a subject of research and under continuous investigation.

After the particle removal stage, the flue gas can be further treated for moisture removal. During this stage, the water vapour and other forms of moisture that exist within the flue gas are removed via cooling and condensation. It should be noted that the use of dry coal fuel reduces the amount of moisture and water content in the flue gas. Flue gas sulphur removal follows the condensation stage so as to ensure that there is no increased risk of corrosion in the boiler. The process is again followed by another stage of cooling and condensation (a gas drier can also be employed after the condensator for dehydrating the remaining water in the flue gas). Finally, and prior to the final stage of CO_2 compression and transportation, removal of the non-condensable gases is necessary. This can be achieved via a super-cooling of the flue gas so as to transfer it in liquid state. Thereafter, the non-condensable gases can be flashed from the liquid CO_2.

Since oxyfuel technology is still at the research and development stage, no operating pilot plants have yet materialized. However, the promising preliminary results of this technology in terms of CO_2 capture have caused significant interest, and a number of oxyfuel plants are already planned to be commissioned within the next 2-3 years.

The necessity of mixing the pure oxygen with recycled CO_2 in the boiler has already been mentioned above. Extensive research is taking place in order to minimize the concentration of recycled CO_2 in the mixture so as to increase oxyfuel plant efficiency and reduce boiler size. Circulating fluidisized bed boilers can be shown to significantly reduce the amount of flue gas recycle due to the fact that the combustion temperature can be controlled through the internal recirculation of bed material instead of the recirculation of CO_2. This is something that cannot be achieved with pulverized fuel boilers used in the conventional coal-fired plants. In a CFB boiler scenario, oxygen concentration in the O_2/CO_2 mixture can become very high (oxygen concentration of up to 70% has been reported). Another method for increasing oxyfuel plant efficiency is the substitution of the ASU by other technologies for pure oxygen provision that are less demanding in terms of auxiliary power requirements. Indeed, use of the ASU has been reported to penalize plant efficiency by as much as 10 percentage points (compared to a conventional pulverized coal plant), while the use of alternative technology for oxygen provision, such as ion-transport membranes may reduce this to only 3-5 percentage points. Apart from the these, other oxygen provision technologies are currently the ceramic auto-thermal recovery and the chemical looping combustion. However, the cryogenic ASU process remains the only available large-scale technology for oxygen provision at present.

3.5. Comparison of CO_2 Capture Technologies

As a general rule, it is acknowledged that CO_2 emissions from an IGCC plant are comparable to those from a pulverized coal plant. In more detail, based on this analysis, in the case of no CCS integration the CO_2 emissions indicator of the IGCC plant is 896g/kWh whereas for the pulverized coal plant is 825g/kWh, meaning that for the same power capacity, the emissions of an IGCC plant are higher by 8%. This is shown in Figure 49, where it is also clear that the CO_2 emissions from a natural gas combined cycle of the same power capacity are 403g/kWh, that is, less than half of the emissions of the IGCC and pulverized coal plants. Also, CO_2 emissions indicator of the oxyfuel plant is much smaller than the emission indicator of the plants with no CSS integration. Specifically, the CO_2 emissions indicator for oxyfuel is 99g/kWh whereas for the pulverized coal plant is 825g/kWh and for the natural gas combined cycle of the same power capacity 403g/kWh, that is, less than half of the emissions of the pulverized coal plant.

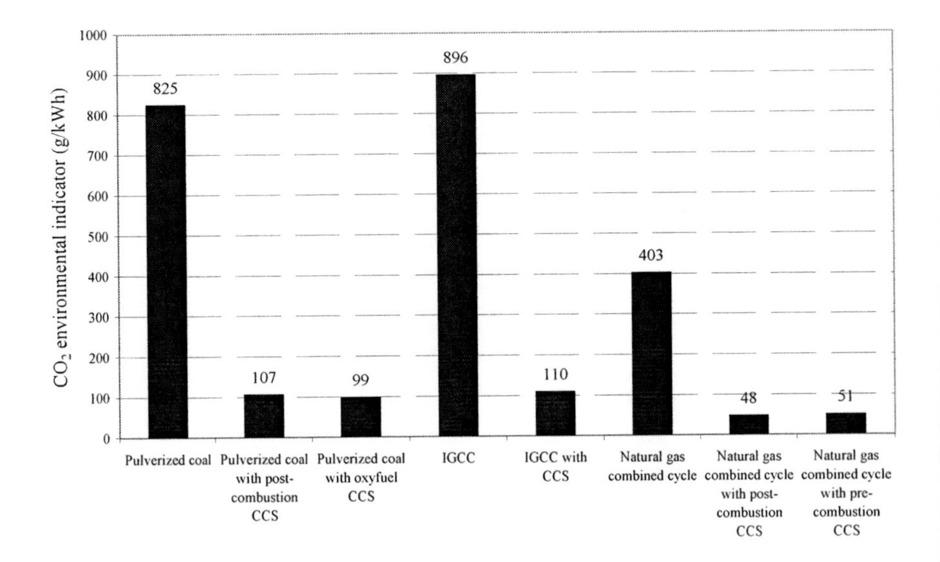

Figure 49. CO_2 emissions reductions from various CCS technologies.

The oxyfuel plant retains the lower CO_2 emissions indicator amongst the other CCS plants that use coal as fuel. Specifically, the CO_2 emissions indicator of the pulverized coal plant (post combustion CCS) is 107g/kWh and of the IGCC plant (pre combustion CCS) is 110g/kWh. However, the CO_2 emissions from a natural gas combined cycle with pre-combustion CCS is 51g/kWh and with post-combustion CCS is 48g/kWh. This is due to the much lower CO_2 emissions inherent in the use of natural gas as combustion fuel.

Chapter 4

4. DIRECT SOLAR RES TECHNOLOGIES

Energy from the sun is called solar energy. The heat that builds up in a car when it is parked in the sun is an example of solar energy. Solar power is the most direct RES technology. This energy comes from processes called solar heating, solar home heating, solar dryer, solar cooker, solar water heating, solar photovoltaic (converting sunlight directly into electricity), and solar thermal electric power (when the sun's energy is concentrated to heat water and produce steam, which is used to produce electricity).

Solar energy is the energy force that sustains life on earth for all plants, animals and people. The earth receives this energy from the sun in the form of electromagnetic waves, which the sun continually emits into space. The earth can be seen as a huge solar energy collector receiving large quantities of this energy which takes various forms, such as direct sunlight, heated air masses causing wind, and evaporation of the oceans resulting as rain which can form rivers. This solar potential can be trapped directly as solar energy, such as solar thermal and photovoltaic systems, and indirectly, such as, wind, biomass and hydroelectric energy.

Solar energy is a renewable source that is inexhaustible and is locally available. It is a clean energy source that allows for local energy independence. The sun's power that is reaching the earth annually is typically about $1000 W/m^2$, although availability varies with location and time of year. Capturing solar energy typically requires equipment with a relatively high initial capital cost. However, in some cases, over the lifetime of the solar equipment, these systems can prove to be cost competitive, as compared to conventional energy technologies.

The solar energy industry is divided into mainly two markets, the photovoltaic (PV) market and the solar thermal market. The solar thermal technology uses the heat radiated from the sun, for purposes such as heating water or power generation. On the other hand PV solar cells use the properties of particular semiconducting materials to convert sunlight energy to electricity.

4.1. The Photovoltaic Technology

Photovoltaic (PV) systems convert energy from the sun directly into electricity. They are composed of photovoltaic cells, usually a thin wafer or strip of semiconductor material that generates a small current when sunlight strikes them. Multiple cells can be assembled into modules that can be wired in an array of any size. Small PV arrays are found in wristwatches and calculators, the largest arrays have capacities in excess of 20MW.

In recent years the PV industry has been experiencing a dramatic growth at a global level. Continuous increase of conventional fuel costs as well as growing pressure to turn towards RES are the main drivers behind this rapidly expanding industry which since the start of the decade has achieved continuous annual growth of around 30%. At a global energy output level, the PV industry is still lagging behind other RES technologies, such as, hydropower and wind energy. This is due to the high costs associated with the manufacturing of PV solar modules, costs that will however steadily diminish as a result of continuous advancements in technology.

PV systems are cost-effective in small stand alone (off-grid) applications, providing power, for example, to rural homes in developing countries, off-grid cottages and motor homes in industrialized countries, and remote telecommunications, monitoring and control systems worldwide. Water pumping is also a notable off-grid application of PV systems that are used for domestic water supplies, agriculture and, in developing countries, provision of water to villages. These power systems are relatively simple, modular, and highly reliable due to the lack of moving parts. PV systems can be combined with fossil fuel-driven generators in periods of little sunshine (e.g., winter at high latitudes) to form hybrid systems.

PV systems can also be tied to isolate or central grids via a specially configured inverter. Unfortunately, without subsidies, grid connected (on-grid) applications are rarely cost-effective due to the high price of PV modules, even if it has declined steadily since 1985. Due to the minimal maintenance of PV systems and the absence of real benefits of economies of scale during construction, distributed generation is the path of choice for future cost-effective on-grid applications. In distributed electricity generation, small PV systems would be widely scattered around the grid, mounted on buildings and other structures and thus not incurring the costs of land rent or purchase. Such applications have been facilitated by the development of technologies and practices for the integration of PV systems into the building envelope, which offset the cost of conventional material and/or labor costs that would have otherwise been spent.

The residential rooftop application of PVs is expected to provide the major application of the coming decade and to provide the market growth needed to reduce prices. Large centralized solar PV power stations able to provide low-cost electricity on a large scale would become increasingly attractive approaching 2020. Solar electricity produced by PV solar cells is one of the most promising options yet identified for sustainably providing the world's future energy requirements. Although the technology has, in the past, been based on the same silicon wafers as used in microelectronics, a transition is in progress to a second generation of a potentially much lower-cost thin-film technology. Cost reductions from both increased manufacturing volume and such improved technology are expected to continue to drive down PV cell prices over the coming two decades to a level where the cells can provide competitively priced electricity on a large scale.

4.1.1. Basics of PVs

Electricity can be produced from sunlight through a process called "photovoltaics", which can be applied, in either a centralized or decentralized way. "Photo" refers to light and "voltaic" to electrical voltage. The term describes a solid-state electronic cell that produces direct current (DC) electrical energy from the radiant energy of the sun.

The basic steps from the PV solar cell to a fully operating PV system are presented in Figure 50. PV solar cells are made of semi-conducting material, most commonly silicon, coated with special additives. When light strikes the cell, electrons are knocked and become loose from the silicon atoms and flow in an in-built circuit producing electricity. Individual solar cells can be connected in series and in parallel to obtain desired voltages and currents. These groups of cells are packaged into standard modules that protect the cells from the environment. PV modules are extremely reliable since they are solid state and there are no moving parts.

PV systems are made up of a variety of components, which aside from the modules, may include conductors, fuses, batteries, inverters, etc. Components will vary, however, depending on the application. PV systems are modular by nature, meaning that systems can be expanded and components easily repaired or replaced if needed. PV systems are cost effective for many remote power applications, as well as for small stand-alone power applications in proximity to the existing electricity grid.

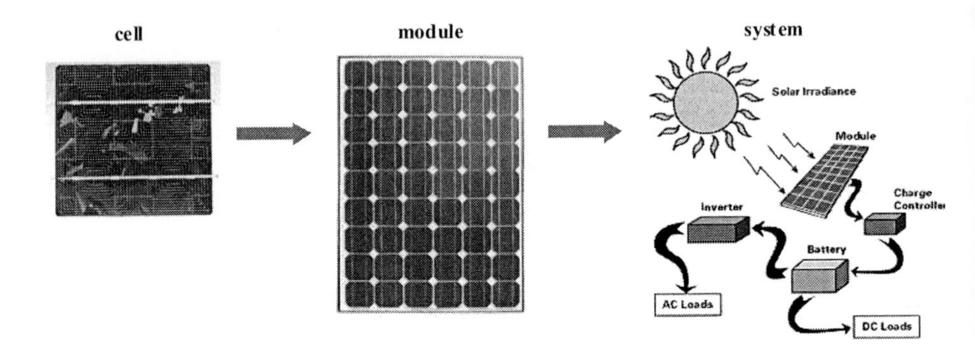

Figure 50. Basic steps from PV cell to PV system.

4.1.2. Principle of Operation

A cell is created when a positively charged (p-type) layer of silicon is placed against a negatively charged (n-type) to create a diode for the flow of electrons. When silicon is exposed to light, electrical charges are generated. Referring to Figure 51, light entering the cell through the gaps between the top contacts metal gives up its energy by temporarily releasing electrons from the covalent bonds holding the semiconductor together; at least this is what happens for those photons with sufficient energy. The p-n junction within the cell ensures that the now mobile charge carriers of the same polarity all move off in the same direction. These electrical charges are conducted away as DC power by placing metal contacts on the top and bottom of a PV cell. If an electrical load is connected between the top and rear contacts to the cell, electrons will complete the circuit through this load, constituting an electrical current in it. Energy in the incoming sunlight is thereby converted into electrical energy.

The cell operates as a "quantum device", exchanging photons for electrons. Ideally, each photon of sufficient energy striking the cell causes one electron to flow through the load. In practice, this ideal is seldom reached. Some of the incoming photons are rejected from the cell or get absorbed by the metal contacts (where they give up their energy as heat). Some of the electrons excited by the photons relax back to their bound state before reaching the cell contacts and thereby the load.

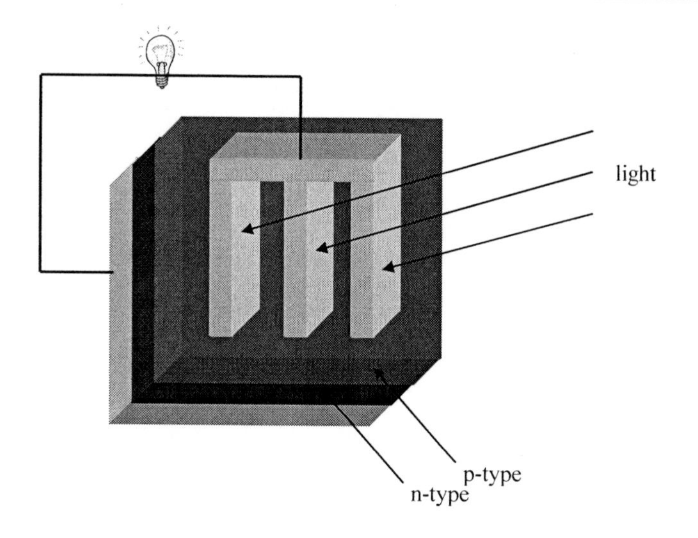

Figure 51. Solar cell operating principle.

The electrical power consumed by the load is the product of the electrical current supplied by the cell and the voltage across it. Each cell can supply current at a voltage between 0.5V-1V, depending on the particular semiconductor used for the cell. Since the electrical output of a single cell is quite small to generate sufficient amount of electricity multiple cells are connected together to form a module or a panel. The PV module is the primary component of PV system and any number of PV modules can be connected to generate the desired amount of electricity. The modular structure of a PV system is considered an advantage since at any instant a new module can be incorporated to the system to satisfy the new electricity requirement. Different PV modules vary in structure; however, they generally include the following elements: (a) glass cover in which the transparent glass cover is placed over the PV cell for protection reasons, (b) anti-reflective sheet which is used to enhance the effect of the glass cover while the anti-reflective coating is used to block reflection, (c) cell and (d) frame and panel backing which are used to hold all the pieces together and protect the PV cell from damage.

4.1.3. PV Technologies

The choice of the semiconductor defines the PV technology. There are two main PV technologies, such as, crystalline silicon solar cells and thin film solar cells as described in the following sections.

4.1.3.1. Crystalline Silicon Solar Cells

The technology used to make most of the crystalline silicon solar cells, fabricated so far, borrows heavily from the microelectronics industry and is known as silicon wafer technology. The silicon source material is extracted from quartz, although sand would also be a suitable material. The silicon is then refined to very high purity and melted. From the melt, a large cylindrical single crystal is drawn. The crystal, or "ingot", is then sliced into circular wafers, less than 0.5mm thick, like slicing bread from a loaf. Sometimes this cylindrical ingot is "squared-off" before slicing so the wafers have a "quasi-square" shape that allows processed cells to be stacked more closely side-by-side. Most of this technology is identical to that used in the much larger microelectronics industry, benefiting from the corresponding economies of scale. Since good cells can be made from material of lower quality than that used in microelectronics, additional economies are obtained by using off-specification silicon and off-specification silicon wafers from this industry.

The first step in processing a wafer into a cell is to etch the wafer surface with chemicals to remove damage from the slicing step. The surface of crystalline wafers is then etched again using a chemical that etches at different rates in different directions through the silicon crystal. This leaves features on the surface, with the silicon structure that remains determined by crystal directions that etch very slowly. The p-n junction is then formed. The impurity required to give p-type properties (usually boron) is introduced during crystal growth, so it is already in the wafer. The n-type impurity (usually phosphorus) is now allowed to seep into the wafer surface by heating the wafer in the presence of a phosphorus source.

Crystalline silicon solar cells hold 93% of the market. Despite the fact that it is a relatively poor light absorbing semiconducting material, over the years it has been the primary raw material used in most solar PV cells due to its ability to yield stable and efficient cells, with efficiencies between 11-16% in terms of converting sunlight energy to electrical energy.

There are two types of crystalline silicon solar cells that are used in the industry, such as, (a) monocrystalline silicon cells (single-Si) and (b) multicrystalline silicon cells (multi-Si or poly-Si).

The monocrystalline silicon cell is made using cells saw-cut from a single cylindrical crystal of silicon. The main advantage of the monocrystalline silicon cells is the high efficiency which is around 15%. The multicrystalline silicon cell is made by sawing a cast block of silicon first into bars and then into wafers. Multicrystalline cells are cheaper to manufacture than monocrystalline ones due to the simpler manufacturing process. However they are slightly less efficient than the monocrystalline with average efficiency of approximately 12%.

In general crystalline silicon solar cells have a relatively high production cost and subsequently high selling price. Moreover, its dependence on purified silicon as the key raw material creates additional difficulty since there is global shortage of the material. The relative high costs result from the complex and numerous production steps involved in wafer and cell manufacturing and the large amount of highly purified silicon feedstock required.

4.1.3.2. Thin-Film Solar Cells

Due to the high production cost of the crystalline silicon wafers, the PV industry has been seeking for alternative ways of manufacturing PV solar cells using cheaper materials such as the thin-film solar cells.

In the thin-film technology approach, thin layers of semiconductor material are deposited onto a supporting substrate, or superstrate, such as a large sheet of glass as indicated in Figure 52. Typically, less than a micron (μm) thickness of semiconductor material is required, 100-1000 times less than the thickness of silicon wafer. Reduced material use with associated reduced costs is a key advantage. Another advantage is that the unit of production, instead of being a relatively small silicon wafer, becomes much larger, for example, as large as a conveniently handled sheet of glass might be. This reduces manufacturing costs.

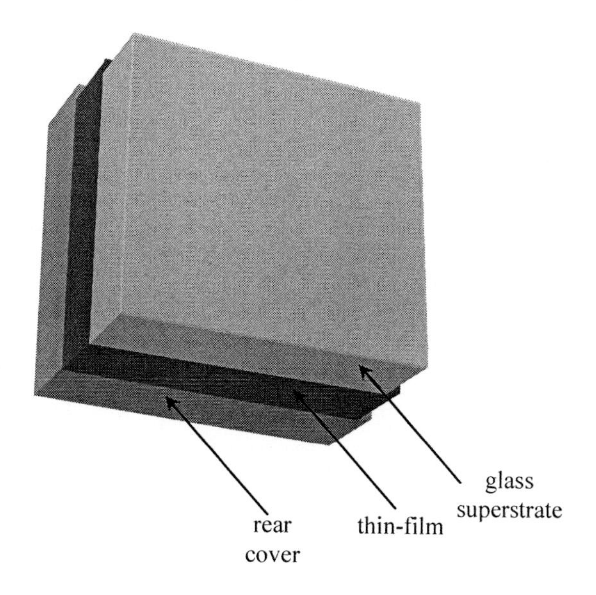

Figure 52. Thin-film technology approach.

Silicon is one of the few semiconductors inexpensive enough to be used to make solar cells from self-supporting wafers. However, in thin-film form, due to the reduced material requirements, virtually any semiconductor can be used. Since semiconductors can be formed not only by elemental atoms, such as silicon, but also from compounds and alloys involving multiple elements, there is essentially an infinite number of semiconductors from which to choose.

At present, solar cells made from different thin-film technologies are either available commercially, or close to being so, such as, cadmium telluride (CdTe), copper indium diselenide (CIS), amorphous silicon (a-Si) and thin-film silicon (thin film-Si). Amorphous silicon is in commercial production while the other three technologies are slowly reaching the market. Over the coming decade, one of the above technologies is expected to establish its superiority and attract investment in major manufacturing facilities that will sustain the downward pressure on cell prices. As each of these thin-film technologies has its own strengths and weaknesses, the likely outcome is not clear at present.

Thin-film panels have several important drawbacks. What they gain in cost savings and flexibility they lose in efficiency resulting in the lowest efficiency of any current PV technology at approximately 6-7%. The main interest in these technologies rises from the fact that they can be manufactured by relatively inexpensive industrial processes, in comparison to crystalline silicon technologies yet they offer typically higher module efficiency than amorphous silicon.

4.1.4. PV Systems

The primary articles of commerce in the PV market are the PV modules which are integrated into systems designed for specific applications. The components added to the module constitute the “balance of system” or BOS. Balance of system components can be classified into four categories (see also Figure 50): (a) batteries which store electricity to provide energy on demand at night or on overcast days, (b) controllers which can manage the energy storage to the battery and deliver power to the load, (c) inverters which are required to convert the DC power produced by the PV module into alternate current (AC) power and (d) structure which is required to mount or install the PV modules and other components.

Not all systems will require all these components. For example in systems where no AC load is present an inverter is not required. For grid connected systems, the utility grid acts as the storage medium and batteries are not required. Some systems also require other components which are not strictly related to PVs.

Some stand alone systems, for example, include a fossil fuel generator that provides electricity when the batteries become depleted; and water pumping systems require a DC or AC pump.

4.1.4.1. PV Modules

To make modules, PV solar cell manufacturers assemble the cells into modules or sell them to module manufacturers for assembly. Because the first important applications of PV involved battery charging, most modules in the market are designed to deliver DC power at slightly over 12V. A typical crystalline silicon PV solar module consists of a series circuit of 36 cells, encapsulated in a glass and plastic package for protection from the environment. This package is framed and provided with an electrical connection enclosure, or junction box.

PV modules are rated on the basis of the power delivered under Standard Testing Conditions (STC) of 1kW/m² of sunlight and a PV cell temperature of 25°C. Their output measured under STC is expressed in terms of "peak Watt" or Wp nominal capacity. For example an annual industry shipments of 165MWp indicates that PV manufacturers made modules with the ability to generate 165MWp of electric power (nameplate capacity) under STC of 1kW/m² of sunlight and 25°C cell temperature. Typical conversion (solar energy to electrical energy) efficiencies for various PV technologies are tabulated in Table 9.

Table 9. Comparison of various PV technologies

Technology	Typical efficiency*	Maximum efficiency* obtained (laboratory)	Cost
single-Si	12-16%	24.0%	Expensive
poly-Si	11-14%	18.6%	Less expensive
a-Si	6-7%	12.7%	Reduced cost

*solar energy to electrical energy conversion efficiency.

4.1.4.2. Batteries

If an off-grid PV system must provide energy on demand rather than only when the sun is shining, a battery is required as an energy storage device. The most common battery types are lead-calcium and lead-antimony. Nickel-cadmium batteries can also be used, in particular when the battery is subject to a wide range of temperatures. Because of the variable nature of solar radiation, batteries must be able to go through many cycles of charge and discharge without damage. The amount of battery capacity that can be discharged without damaging the battery depends on the battery type. Lead-calcium batteries are suitable only in "shallow cycle" applications where less than 20% discharge occurs each cycle. Nickel-cadmium batteries and some lead-antimony batteries can be used in "deep cycle" applications where the depth of discharge can exceed 80%.

Batteries are characterized by their voltage, which for most applications is a multiple of 12V, and their capacity, expressed in Ampere-hours (Ah). For example a 50Ah, 48V battery will store 50Ah × 48V = 2,400Wh of electricity under nominal conditions. Note that optimizing battery size is critical in obtaining good battery life, suitable system performance, and optimal system life-cycle costs. Unnecessary battery replacement is costly, particularly for remote applications.

4.1.4.3. Power Conditioning

Several electronic devices are used to control and modify the electrical power produced by the PV array. These include (a) battery charge controllers, which regulate the charge and discharge cycles of the battery, (b) maximum power point trackers (MPPT), which maintain the operating voltage of the array to a value that maximizes array output, (c) inverters, in order to convert the DC output of the array or the battery into AC (AC is required by many appliances and motors, it is also the type of power used by utility grids and therefore grid connected systems always require the use of an inverter) and (d) rectifiers (battery chargers) in order to convert the AC current produced by a generator into the DC current needed to charge the batteries.

4.1.4.4. Generators

For off-grid applications it is also possible to have both a PV system and a fossil fuel generator running in parallel. The use of a generator eliminates the need to oversize the PV array and the battery bank in order to provide power during periods with little sunshine. The PV array and the generator supplement each other, the PV array reduces the fuel use and maintenance cost of the generator and the generator replaces the part of the PV system that would need to be oversized to ensure an uninterrupted supply of power.

Generators can use a variety of fossil fuels, such as gasoline, diesel oil, propane or natural gas. The requirement for a generator, and the fraction of the load met respectively by the PV system and the generator, will depend on many factors, including the capital cost of the PV array, operating costs of the generator, system reliability, and environmental considerations (e.g., noise of the generator, emission of fumes, etc.).

4.1.5. PV Power Systems

The PV power systems are classified into categories according to their operational requirements, their component configurations and how the system is connected to other power sources and loads. The two principal PV systems described in the following sections are the grid connected (or on-grid) systems and the stand alone systems (or off-grid) systems.

4.1.5.1. Grid Connected Systems

In grid connected applications the PV system feeds electrical energy directly into the electric utility grid (this includes central grids and isolated grids). Two application types can be distinguished, distributed generation and central power plant generation. An example of a distributed grid connected application is building integrated PV for individual residences or commercial buildings. The system size for residences is typically in the 2kWp to 4kWp range. For commercial buildings, the system size can range up to 100kWp or more. Batteries are not necessary when the system is grid connected.

In grid-connected systems, the PV system is designed to operate in parallel with the electric utility grid. The primary component of a grid connected system is the inverter which is used to convert the DC electrical supply to AC electrical supply. In some cases a bi-directional interface can be made between the PV system AC output circuit and the utility network. This will allow the PV system to either supply on-site electrical loads or feed back to the grid electricity excess of the load demand.

The benefits of grid connected PV power generation are generally evaluated based on its potential to reduce costs for energy production and generator capacity, as well as its environmental benefits. For distributed generation, the electric generators (PV or other) are located at or near the site of electrical consumption. This helps reduce both energy (kWh) and capacity (kW) losses in the utility distribution network. In addition, the utility can avoid or delay upgrades to the transmission and distribution network where the average daily output of the PV system corresponds with the utility's peak demand period (e.g., afternoon peak demand during summer months due to air conditioning loads). PV manufacturers are also developing PV modules which can be incorporated into buildings as standard building components such as roofing tiles and curtain walls.

This helps reduce the relative cost of the PV power system by the cost of the conventional building materials, and allows the utility and/or building owner to capture distributed generation benefits. The use of PV in the built environment is expanding with demonstration projects in industrialized countries. Central PV power generation applications are not currently cost-competitive and strong subsidies are required. Several multimegawatt central PV generation systems have however been installed as demonstration projects, designed to help acquire experience in the management of central PV power plants. Installations of central PV generation, like distributed grid connected PV, represent a long-term strategy by governments and utilities to support the development of PV as a clean energy.

4.1.5.2. Stand Alone Systems

Currently, PV is most competitive in isolated sites, away from the electric grid and requiring relatively small amounts of power, typically less than 10kWp. In these off-grid applications, PV is frequently used in the charging of batteries, thus storing the electrical energy produced by the modules and providing the user with electrical energy on demand. Stand alone systems are designed to work independent of the electric utility grid, and are designed and sized to supply certain DC and/or AC loads. These stand alone systems may be powered by a PV array only, or it may be powered by a PV-hybrid system which combines either wind and solar energy, or an engine-generator with solar energy. In many stand alone systems a battery is used for energy storage to provide electricity support during night time when there is no sunlight available.

The key competitive arena for PV in remote off-grid power applications is against electric grid extension; primary (disposable) batteries; or diesel, gasoline and thermoelectric generators. PV competes particularly well against grid extension for small loads, far from the utility grid. Compared to fossil fuel generators and primary batteries, the key advantage of PV is the reduction in operation, maintenance and replacement costs; these often result in lower life-cycle costs for PV systems.

4.1.5. Summary on PVs

During the past 20 years there has been a significant growth of the solar PV electric technology and governments in Japan, USA and Europe who are the major players of the solar energy market, are all providing financial incentives in order to encourage their people to adopt the eco-friendly PV systems as an alternative source of energy. The use of PV systems provides a number of key benefits:

- Energy security: Solar energy provides reliable access to energy where it is used. It can also supplement energy needs during blackouts and disaster recovery for electricity, water pumping and hot water,
- Energy independence: Solar energy can be used to reduce our independence on fossil fuels imported from foreign countries,
- Eco-friendly: Solar energy is a non polluting source of energy. The significant adaptation to PV electricity could further reduce CO2 emissions in the environment,
- Economic benefits: When installed properly in homes, businesses it can begin to save money immediately,
- Job creation: New jobs are created in manufacturing, distribution, and also many building related jobs for electricians, plumbers, roofers, designers and engineers.

PV is a relatively new technology, which offers a new vision for consumers and business as to how power can be provided alternatively. PV technology is already proving to be a force for social change in rural areas in less developed countries. The unique characteristic of PV is that it is a "radical" and "disruptive" type of technology as compared to conventional power generation technologies.

An overall comparison of the efficiency and the cost of the various PV solar cell technologies are tabulated in Table 9. In general as the efficiency of PV solar cells increases their cost increases too. Also, the technological developments expected over the coming decades are presented in Table 10.

The photovoltaics advantages can be summarized as:

- no fuel is used,
- complementarily with other energy sources, both conventional and renewable,
- flexibility in terms of implementation since PV systems can be integrated into buildings, installed as separate mobile or non-mobile modules, or in central electricity generating stations,
- no wastes or greenhouse gases are produced and
- no moving parts, therefore, almost no maintenance.

The photovoltaics disadvantages can be summarized as:

- low efficiency (around 14%) and
- high capital cost.

Current research and development in the field of photovoltaics includes:

- the improvement of the efficiency to reach 25%,
- the reduction of electricity generation to 5-12€c/kWh,
- the increase of the technical lifetime to reach 40 years and
- the development of advance balancing and storage technologies for large scale implementation of PVs.

Table 10. Expected development of PV technology over the coming decades

Item	1980	Today	2015-2020	2030	Long term potential
Typical turn-key system price (€/Wp, excl. VAT)	>30	5	2.5-2.0	1	0.5
Typical electricity generation costs southern Europe (€/kWh)	>2	0.30	0.15-0.12 (competitive with retail electricity)	0.06 (competitive with wholesale electricity)	0.03
Typical commercial module flat plate* efficiencies	up to 8%	up to 15%	up to 20%	up to 25%	up to 40%
Typical commercial concentrator** module efficiencies	~10%	up to 25%	up to 30%	up to 40%	up to 60%

* Flat plate refers to standard modules for use under natural sunlight.

** Concentrator refers to systems that concentrate sunlight (and, by necessity, track the sun across the sky).

4.2. Solar Thermal

Solar thermal power generation utilizes the sun as a source of heat which can be exploited by concentrating that heat and using it to drive a heat engine to produce power. As such, solar thermal power generation is much more closely related to traditional forms of power generation based on fossil fuel combustion which also rely on heat engines to convert heat into electrical energy.

Solar thermal generation is not new. The first patent for a solar collector was granted in Germany in 1907. However, the first major effort to exploit the sun as a heat source for power generation began in the US after the oil crises of the 1970s and the first commercial plants appeared in the late 1980s in California. Funding for development and deployment of solar thermal generation tailed off soon afterwards when cheap natural gas dominated the power generation market in most parts of the developed world. However, the combination of global warming and volatile gas prices has had a potent effect and both interest and investment in solar thermal power technology are now accelerating rapidly.

Investment is what solar thermal power technology has lacked for most of the past twenty years. With sufficient investment there is no doubt that solar thermal electricity generation can provide an economical source of electricity. The conditions now seem right for it to prosper. With several major projects proposed, under construction or recently entering service there is finally a strong chance that this electricity generating technology can become a part of the main stream, alongside wind, hydro and solar photovoltaic technologies, as a key source of renewable energy for the future.

4.2.1. Available Solar Thermal Power Technologies

Current solar thermal power technologies are distinguished in the way they concentrate solar radiation, such as, (a) parabolic trough systems, (b) solar tower systems and (c) solar dish systems. The direct radiation is concentrated using reflectors and the energy concentrated in this way is transformed into steam, which is used to drive conventional electricity generators. A brief description of each most common available solar thermal power technology and of thermal storage is provided below.

4.2.1.1. Parabolic Trough Technology

A parabolic trough is a long, trough-shaped reflector with a parabolic cross-section as indicated in Figure 53. As a result of this cross-section, sunlight reflected within the trough is focused along a line running the length of the trough. In order to collect this heat, a pipe is positioned along the length of the trough at its focus and a heat collection fluid is pumped through it. The tube (or receiver) is designed to be able to absorb most of the energy focused onto it and must be able to withstand the resultant high temperature.

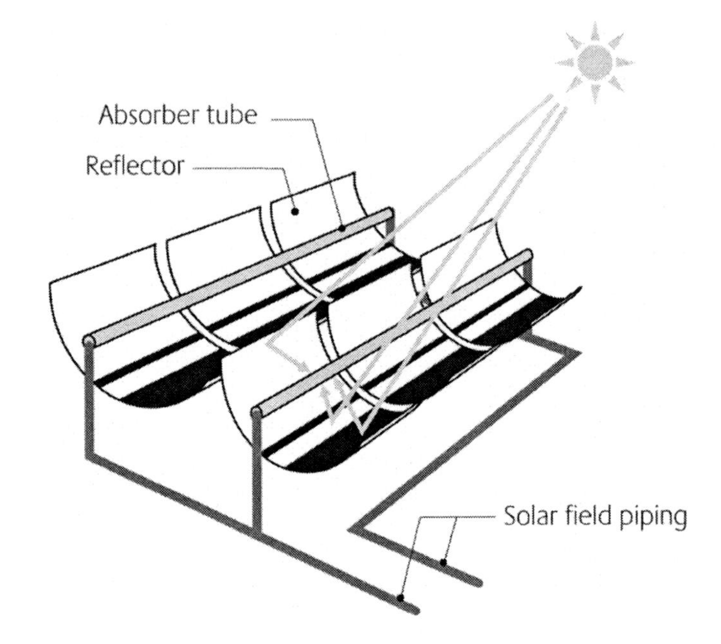

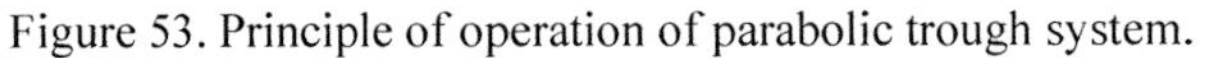
Figure 53. Principle of operation of parabolic trough system.

Typical receivers for this purpose are made of steel tubing with a black coating and surrounded by a protective glass cover with the space between the two evacuated to reduce heat loss. An anti-reflective coating may be added to the outer glass surface to increase efficiency further.

The solar array of a parabolic trough power plant consists of several parallel rows of parabolic reflectors. The heat collecting fluid which is pumped through the pipes along the length of each solar trough is typically synthetic oil, similar to engine oil, capable of operating at high temperature. During operation it is likely to reach between 300°C and 400°C. After circulating through the receivers the oil is passed through a heat exchanger where the heat it contains is extracted to raise steam in a separate sealed system and the steam is then used to drive a steam turbine generator to produce electricity. The heat collecting fluid is then cycled back through the solar collector field to collect more heat.The parabolic troughs along which these tubular receivers run may be five to six meters wide, one or two meters deep and up to 150m in length (though an individual trough of this length will usually be constructed from modular sections). Many of these are required to collect sufficient energy to provide heat for a single power plant. As a consequence, these solar troughs form a physically large part of the solar plant and their cost can have a significant impact on plant economics.

Parabolic solar troughs are usually aligned with their long axes from north to south and they are mounted on supports that allow them to track the sun from east to west across the sky. These supports may be made of steel or aluminum. In the first commercial plants the actual mirrors were made from 4mm glass which is both heavy and expensive. Modern developments aim to reduce the cost and weight by using new techniques and materials including polished aluminum instead of coated glass mirrors. Energy conversion efficiency is one of the keys to commercial success for solar thermal plants. The reflecting mirrors must be both accurately shaped, and accurately positioned in order to achieve maximum solar collection efficiency. Then the tracking system must ensure that each trough is in the optimum position, all day. Finally the tubular energy receivers must operate at the highest efficiency possible too.

The heat collection system currently employed in parabolic trough plants involves use of synthetic oil as the heat collection fluid and, as outlined above, this must be passed through a heat exchanger in order to raise steam to drive a turbine. An obvious simplification to this system can be achieved by replacing the heat collection fluid with water. If water is used, it can be converted directly to steam within the heat collection pipes of the collector field and then used immediately to drive the plant's steam turbine without the need for heat exchangers. This could result in a significant reduction in plant costs but technical requirements are more demanding.

With solar radiation only available for part of each 24 hours, energy storage, as discussed above, represents one means of providing power around the clock. An alternative way of exploiting solar energy when continuous power is needed is to combine a solar thermal power plant with a fossil fuel power plant. Solar heat, when available, can then be used to supplement the heat available from combustion of fossil fuel, reducing carbon dioxide emissions and increasing the renewable contribution to gross power generation.

The most promising hybrid arrangement under consideration involves building a solar thermal power plant based on parabolic troughs alongside a combined cycle power plant, an arrangement called an integrated solar combined cycle (ISCC) plant. In an ISCC plant the combined cycle plant is built in a standard configuration with a gas turbine burning natural gas to generate electricity while the exhaust heat from the turbine is fed into a waste heat boiler, generating steam to drive a steam generator. In the hybrid plant, however, the heat from the solar collectors is used to supplement the heat from the gas turbine exhaust, increasing the output from the steam turbine section of the plant. Currently such plants are under construction in Morocco, Algeria and Egypt.

4.2.1.2. Solar Towers

Solar towers (often called solar central receiver power plants) offer an alternative method of exploiting the energy from the sun in a solar thermal power plant. In this case the collector field consists of an array of heliostats (mirrors) at the centre of which is a tower as illustrated in Figure 54. At the top of the tower is a receiver designed to collect the heat from the sun.

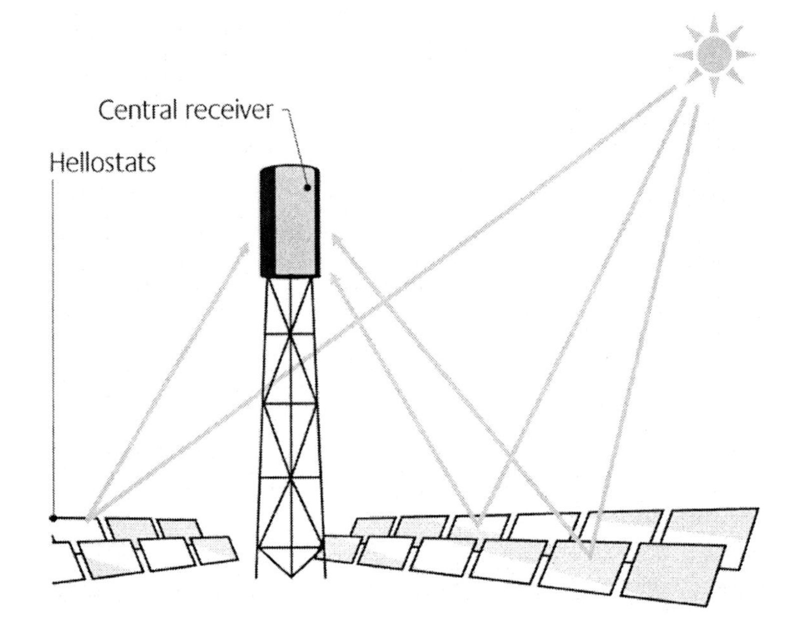

Figure 54. Principle of operation of solar tower system.

In operation each heliostat has an individual tracking system and all are aligned so that the sunlight striking them is directed onto the receiver which is located at the top of the central tower. As the sun moves across the sky, each mirror must be moved too if high collection efficiency is to be maintained. The receiver itself is designed to absorb the energy from the sunlight incident upon it and transfer it to a heat transfer fluid. Depending on system design, this heat transfer may be either, water, a molten salt or air. Solar towers are normally designed with energy storage capability so that they can, in principle, operate 24 hours a day.

4.2.1.3. Solar Dishes

A solar dish power plant uses a circular parabolic dish to collect solar radiation and bring it to a focus, as illustrated in Figure 55. A heat engine situated at the focus exploits the heat generated by this concentration to provide mechanical motion which drives a generator. In the case of the solar dish, the heat engine is normally a special type of engine called a Stirling engine which has extremely high efficiency. There have also been attempts to use small gas turbines based on the Brayton cycle (the thermodynamic cycle upon which the gas turbine is based).

Typical dishes are between five and ten meters in diameter and with reflective areas of $40m^2$ to $120m^2$ though they have been built as large as $400m^2$. Material limitations are likely to restrict the practical size of dishes though dishes up to 15m in diameter ($700m^2$) have been proposed. Dishes in this size range could provide up to 50kW of power. However, today Stirling engines are limited to 25kW. These are best matched with smaller dishes. Gas turbine heat engines based on micro turbines can provide higher output but they are significantly less efficient that Stirling engines. Both micro gas turbine and Stirling engine-based systems can be designed for hybrid operation using a combination of solar heat and the heat from combustion of natural gas.

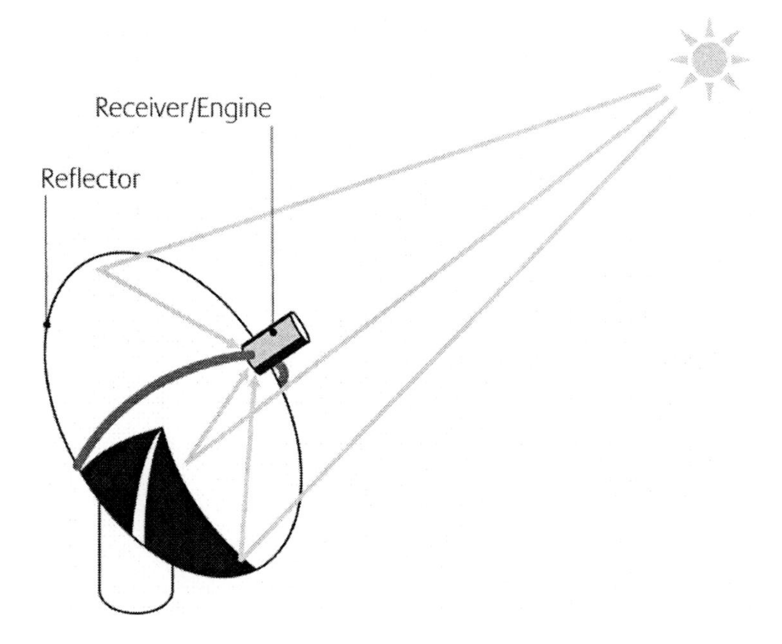

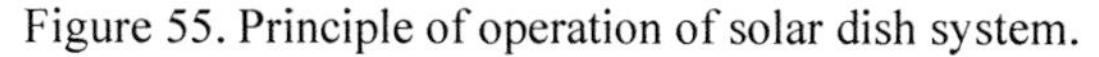
Figure 55. Principle of operation of solar dish system.

As with both parabolic trough collectors and heliostats, solar dishes have to be able to track the sun across the sky in order to achieve maximum efficiency. Tracking systems tend to be expensive and this means that the cost of the dish plays a significant role in the economics of the power system. The dish support is usually constructed as a lattice upon which individual curved mirrors are mounted to create the overall dish. These mirrors may be of glass or polished metal, circular or rectilinear. At the centre of the dish there is a projecting beam to which the heat engine is attached, positioned so as to capture the heat concentrated at the focus of the dish. Dishes with Stirling engines have been built in sizes ranging from 5kW to 25kW. These engines are in theory capable of 40% energy conversion efficiency although practical engines today achieve closer to 30%.

4.2.1.4. Solar Chimney

The solar chimney combines a large collector greenhouse with a central chimney. Hot air is produced by the sun (direct and diffuse radiation) under a large glass roof (collector). The heated air flows to a chimney in the center of the collector and is drawn upwards Figure 56. This updraft drives wind turbines installed at the base of the chimney. The ground under the collector greenhouse functions as a thermal storage. This effect can be increased by covering the ground with black water-filled tubes. Thus, it is possible to operate a solar chimney even at night at a reduced output, i.e., 24 h/day.

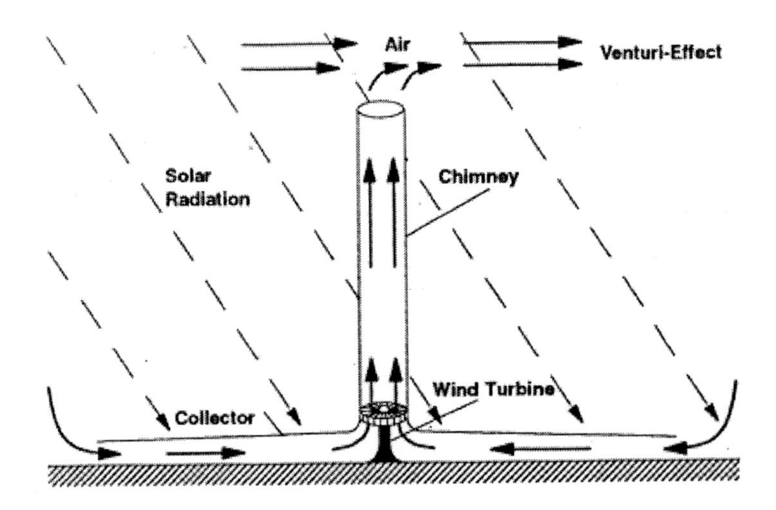

Figure 56. Schematic of a solar chimney.

4.2.2. Thermal Storage

The molten salt system, used for thermal storage in solar thermal power plants, is typically a mixture of sodium and potassium nitrates which melts at about 220°C. In operation the salt is stored in a tank maintained at about 300°C. Molten salt is taken from this tank and passed through the high temperature receiver where it absorbs heat provided by the mirrors from the collector field and is then returned to a high temperature storage tank at a temperature of around 550°C. At this temperature the salt can act as a source of high-grade heat and it appears possible to operate with even higher temperatures if necessary.

Electricity is generated by taking molten salt from the hot storage tank and passing it through a heat exchanger where the heat it contains is transferred to water, generating steam to drive a steam turbine. The cooled molten salt is then returned to the cold storage tank ready to pass through the solar energy receiver once more. By careful sizing of a plant of this type it is quite feasible to build a power station capable of providing power throughout the day and night.

4.2.3. Solar Thermal Power Plants Around the World

Solar thermal power technologies still need further research to overcome non-technical and technical barriers. Solar thermal power plants require a long-term view in the same way as traditional energy producing plants, and therefore benefit from stable policies and continuity of legal and financial frameworks, ideally favorable for solar thermal power systems. Additionally, with little commercial experience to draw on, realistic costs estimates for solar thermal power plants are extremely difficult to make, however, it is expected that cost reduction will result from technical progress. In this section, the available solar thermal power plants in operation and under construction around the world are briefly presented.

4.2.3.1. Solar Thermal Power Plants in Operation

Solar Electric Generating System

Solar Electric Generating System (SEGS) is the largest solar energy generating facility in the world. It consists of nine solar power plants, located at Mohave Desert, in California, USA, with an annual solar potential of 2700kWh/m^2. The plants have a total of 354MW installed capacity, with a gross average output for all nine plants around 75MW. In addition, the turbines can be utilized at night by burning natural gas. The combined solar field has a total parabolic reflecting mirror area of over 2 million m^2 and the nine SEGS plants cover a land area of more than $6,400,000m^2$. Lined up, the parabolic mirrors would extend to over 369km. The SEGS installation uses parabolic trough technology along with natural gas to generate electricity. Around 90% of the electricity is produced by the sunlight. Natural gas is only used when the solar power is insufficient to meet the electricity demand of southern California. The installation uses synthetic oil as heat transfer fluid which heats to over 400°C, transferring its heat to generate steam from water, in order to drive the Rankine cycle steam turbine thereby generating electricity.

Nevada Solar One

Nevada Solar One is a solar thermal plant, based on the parabolic trough technology and is located in the El Dorado Valley in Nevada, USA. The solar field is made up of 760 solar parabolic trough collectors, each with a reflective surface of $470m^2$, to make up a total of $357,200m^2$ of solar reflective field, over a total land area of $1,600,000m^2$. The steam turbine has a nominal generating capacity of 64MW and the plant produces annually around 130GWh (annual capacity factor of 23%), while employing a supplementary gas heater facility for back-up steam generation in case solar irradiation is not adequate.

PS10

PS 10 is an operational solar thermal plant based on the solar tower technology. It is located in Sanlucar de Mayor in Sevilla, Spain and begun operation in 2007. The plant has a land area of $600,000m^2$ and it is the first solar tower plant to begin commercial electricity generation operations in the world. The plant solar tower is 100m high, and the heliostats that track the sun on two axes, and concentrate the sun's irradiation to the focal point (receiver) located on the tower, are 624 in total with a surface area of $120m^2$ each. Therefore, the total reflective surface area is $75,000m^2$. Each heliostat has an independent solar tracking mechanism that directs solar radiation toward the receiver. The actual heliostat field does not however completely surround the receiver tower. In the northern hemisphere, the heliostat field is located on the north side of the tower to optimize the amount of solar radiation collected while minimizing heat loss. The receiver is located in the upper section of the tower. It is a "cavity" receiver and is comprised of four vertical panels that are 5.5m wide and 12m tall. The panels are arranged in a semi-cylindrical configuration and housed in a square opening 11m per each side.

With an annual solar potential of $2100kWh/m^2$, and installed capacity of 11MW, the plant is capable of generating 24.3GWh of electricity annually (annual capacity factor of 25%). PS10 plant is capable of storing 1 hour worth of steam for electricity generation via steam storage tanks. Steam is stored at 50bar and 285°C, and it condenses and flashes back to steam, when the pressure is lowered. Additionally, under low solar irradiation conditions, the plant is capable of supplying 12%-15% of its capacity via natural gas combustion. The total plant efficiency, (conversion of solar irradiation to electricity) is approximately 17%. This is a fairly high number considering that the efficiency of the steam cycle alone is approximately 27%.

Andasol 1 and 2

Andasol 1 and 2 are two identical solar thermal plants expected to begin operations very soon and will then be the first solar thermal parabolic trough power plants to operate in Europe. These two 50MW plants are located in Andalucia, Spain.

The Andasol 1 solar thermal plant consists of three basic parts: the solar field, the storage tanks and the power generation block. The solar field of each of the Andasol plant uses 624 parabolic mirrors arranged in 156 loops with a total reflective area of more than 510,120m^2 in a land area of 2,000,000m^2. Andasol 1 is estimated to supply annual electricity generation of 179GWh. With an annual solar potential of 2201kWh/m^2, total solar field annual average efficiency (efficiency of solar irradiance conversion to solar steam) is estimated around 43%, while the steam cycle efficiency is estimated to be 38.1%. Overall plant efficiency is thus around 16%.

The Andasol plants are the first solar thermal plants to utilize two molten salt storage tanks for heat storage in cases of low solar irradiation. Heat storage begins to occur at midday, when the sun irradiation is very high and electricity can be generated while, at the same time, the heat storage system can be charged. In order to charge the storage system, heat from the heat transfer fluid is transferred to the molten salt tank which collects the heat while the molten salt moves from the cold tank to the hot tank, where it accumulates until it is completely full. When heat is to be discharged, the salt cools down and moves to the cold tank. Since the cold and hot salts are kept in two separate tanks, this is called a two-tank system. The molten salt storage tank system increases the annual equivalent full-load running time of the solar thermal plant to around 3500 hours. The two storage tanks have a diameter of 36m and a height of 14m each and have a storage capacity of 7.5h at 50MW. The quantity of the molten salts employed is estimated to be 28500t, with a melting temperature of 221°C and allowed operational temperature range between 291°C (cold tank) - 384°C (hot tank).

4.2.3.2. Solar Thermal Power Plants under Construction

Solnova 1

Solnova 1 is a solar thermal plant to be manufactured in the Sanlucar de Mayor location in Seville, Spain, and it is based on the parabolic trough technology. The plant will use synthetic oil to generate high temperature steam and run a conventional steam cycle. The plant will be comprised of 90 rows of collectors oriented north-south. Every row will have 4 trough modules (therefore a total of 360 modules) that are each 12.5m long and 5.76m wide. Each module will rotate about its axis to track the sun. Enough space will be left between the rows to reduce losses due to shading and allow for easy operation and maintenance. The total reflective surface will be composed of approximately 260,000m^2 of mirrors. The total land area required for the Solnova 1 plant will be around 1,200,000m^2.

Solnova 1 installed capacity will be 50MW and will be capable of generating 114.6GWh of clean electrical energy annually (annual average capacity factor of 26%). In low solar irradiation conditions, the plant will be capable of supplying 12%-15% of its capacity through natural gas combustion. At peak conditions, the plant will convert solar radiation into heat at efficiency near 57%. Combine with the efficiency of 34% of the steam cycle, the overall plant efficiency is estimated to be approximately 19%.

PS20

PS20 is a solar thermal tower plant under construction, with a similar technology to PS10. It is constructed next to the PS10 plant but with twice the capacity, that is 20MW. PS20 will have 1,255 sun-tracking heliostats each with a surface area of 120m2 and a solar tower 160m long. It is expected to be able to generate 48.6GWh per year, with a total land requirement of 900,000m2. PS20 will also have the possibility to burn natural gas to cover 12-15% of its capacity in case of low solar irradiation.

Solar Tres

Solar Tres solar thermal plant is being constructed in Andalucia, Spain. It is a solar tower based solar thermal plant with a capacity of 19MW. Solar Tres will employ 2,480 heliostats of a total reflective surface area of approximately 300,000m^2 (120m^2 surface area per heliostat), located in a land area of 1420000m^2. Annual solar potential levels reach 2060kWh/m^2. A unique feature of the Solar Tres plant is the use of molten salt as a heat transfer medium in the interior of the receiver, instead of the heat transfer fluid (synthetic oil) normally used in solar thermal power plants. This allows for the collection, transport and storage of the thermal energy with very high efficiencies through the high differential temperatures. Also, the molten salt flow loop reduces the number of valves, eliminates "dead legs" and allows fail-safe draining that keeps salt from freezing. Compared to the Andasol plants, Solar Tres will store 3 times more energy per kg of salt. The concentration of the sunlight on the receiver, situated on the top of a 120m high tower, will produce temperatures of over 850°C and the salt will be heated to approximately 565°C. The salt then flows in molten state through a heat exchanger in which sufficient steam will be produced to operate the steam turbine of the power block. Annual electricity generation is expected to be 96.4GWh, and overall plant efficiency around 14%.

Solar Tres plant will employ a large thermal storage system by using 6250t of salt with insulated storage tank immersion heaters. This high capacity liquid nitrate salt storage system is efficient and low-risk and is designed for a high-temperature liquid salt at 565°C (with a daily temperature drop of only 1-2°C) and a cold temperature salt at 45°C above its melting point (240°C). The storage system may provide back-up steam for up to an additional 15 hours. Solar Tres design will also employ a 43MW steam generator system that will have a forced recirculation steam drum. This innovative design places components in the receiver tower structure at a height above the salt storage tanks that allows the molten salt system to drain back into the tanks, providing a passive fail-safe design. This will improve plant availability and reduced operation and maintenance costs.

Ibersol 1

The Ibersol 1 plant will be situated in Puertollano, Spain and will have a capacity of 50MW, based on a parabolic trough technology. There will be 576 parabolic trough collectors arranged in 216 loops of 4 collectors per loop. Steam generation will be achieved via the use of a heat transfer fluid (thermal oil) and a thermal storage stage will also be constructed, based on the technology of molten nitrate salt tanks.

4.2.4. Overall Comparison

The primary consideration for a solar thermal power plant is the amount of land needed, which is significant. The land area requirements depend on the available area solar potential as well as on the degree of the integrated thermal storage.

Table 11. Indicative land area requirements for solar thermal power plants

Solar thermal power plant	Capacity (MW)	Thermal storage (h)	Land area (m^2)	Specific land area (m^2/kW)
Parabolic trough technology				
SEGS	354	-	6,400,000	18
Nevada Solar One	64	-	1,600,000	25
Andasol	50	7.5	2,000,000	40
Solnova	50	-	1,200,000	24
Solar tower technology				
PS10	11	1	600,000	55
PS20	20	-	900,000	45
Solar Tres	19	15	1420000	75

Also, with little commercial experience to draw on, realistic land estimates for solar thermal power plants are extremely difficult to make. The land area requirements of the various existing solar thermal power plants are tabulated in Table 11. Parabolic troughs require a land area of approximately 25m^2/kW, in the case where no thermal storage is integrated. Solar towers have the highest requirement of approximately 45m^2/kW, in the case where no thermal storage is integrated.

Chapter 5

5. Indirect Solar RES Technologies

Most renewable energy sources on Earth are forms of indirect solar energy, although we usually don't think of them in that way. Coal, oil, and natural gas derive from ancient biological material that took its energy from the Sun (via photosynthesis) millions of years ago. All the energy in wood and foodstuffs also comes from the Sun. Movement of the wind (which causes waves at sea), and the evaporation of water to form rainfall, which accumulates in rivers and lakes, are also powered by the Sun. Therefore, hydroelectric power and wind and wave power are forms of indirect solar energy.

5.1. The Wind Turbine Technology

Winds are caused by the rotation of the earth and heating of the atmosphere by the sun. Due to the heating of the air at equatorial regions, the air becomes lighter and starts to rise, and at the poles the cold air starts sinking. The power in the wind is proportional to the cube of the wind speed. It is, therefore, essential to have detailed knowledge of the wind and its characteristics if the performance of wind turbines is to be estimated accurately.

Wind energy converts the power available in moving air into electricity. Wind turbines turn in by the moving air and drive an electric generator. The generator then supplies the electric current. Wind energy is renewable and environmentally benign. Wind-driven electric generators could be utilized as an independent power source and for purposes of augmenting the electricity supply from grids. Wind energy potential increases very rapidly with increasing wind speed. The annual wind speed at a location is useful as an initial indicator of the value of the wind resource. The indicated wind potential based on mean annual wind speed is indicted in Table 12.

Table 12. Indicated potential based on mean annual wind speed

Annual mean speed at 10m height	Indicated value of wind potential
< 4.5m/s	Poor
4.5m/s – 5.4m/s	Marginal
5.4m/s – 6.7m/s	Good to very good
> 6.7m/s	Exceptional

Large, modern wind turbines operate together in wind farms to produce electricity. Small turbines are used by homeowners and farmers to help meet localized energy needs. Wind turbines capture energy by using propeller-like blades that are mounted on a rotor. These blades are placed on top of high towers, in order to take advantage of the stronger winds at 30 meters or more above the ground. This is illustrated in Figure 57.

Figure 57. The wind turbine technology.

The wind causes the propellers to turn, which then turn the attached shaft to generate electricity. Wind can be used as a stand-alone source of energy or in conjunction with other renewable energy systems.

The first wind turbines for electricity generation had already been developed at the beginning of the twentieth century. The technology was improved step by step since the early 1970s. By the end of the 1990s, wind energy has re-emerged as one of the most important sustainable energy resources. During the last decade of the twentieth century, worldwide wind capacity has doubled approximately every three years. Costs of electricity from wind power have fallen to about one-sixth since the early 1980s. And the trend seems to continue. It is predicted that the cumulative capacity will be growing worldwide by about 25% per year until 2010 and cost will be dropping by an additional 20% to 40% during the same time period.

There are many onshore wind farms around the world. Offshore wind farms in coastal waters are being developed because winds are often stronger blowing across the sea.

Wind energy technology itself also moved very fast towards new dimensions. This is illustrated in Table 13. At the end of 1989, a 300kWe wind turbine with 30 meters rotor diameter was the state of the art. Only 10 years later, 1500kWe turbines with a rotor diameter of around 70 meters are available from many manufacturers. The first demonstration projects using 2MWe wind turbines with a rotor diameter of 74 meters were installed before the turn of the century and are now commercially available. Currently, under development are 7MWe wind turbines and the first prototypes are expected to install soon.

It is important to mention that more than 80% of the world-wide wind capacity is installed in only five countries: Germany, USA, Denmark, India and Spain. Hence, most of the wind energy knowledge is based in these countries. The use of wind energy technology, however, is fast spreading to other areas in the world.

Wind energy was the fastest growing energy technology in the 90s, in terms of percentage of yearly growth of installed capacity per technology source. The growth of wind energy, however, is not evenly distributed around the world. By the end of 1999, around 70% of the world-wide wind energy capacity was installed in Europe, a further 19% in North America and 9% in Asia and the Pacific. Horizontal-axis, medium to large size grid-connected wind turbines with a capacity greater than 100kWe have, currently, the largest market share and it is expected, also, to dominate the development in the near future. Horizontal-axis wind turbines can be designed in three different ways as shown in Table 14.

Table 13. Development of wind turbine size between 1985 and 2002

Year	Capacity (kW)	Rotor diameter (m)
1985	50	15
1989	300	30
1992	500	37
1994	600	46
1998	1500	70
2002	3500-4500	88-120
2005	4000-6000	90-130
2009	7000	140

Table 14. Basic design approaches of horizontal axis wind turbines

Design approach	Technical characteristics
A	Turbines designed to withstand high wind loads Optimize for reliability High solidity but non-optimum blade pitch Three or more blades Lower rotor tip speed ratio
B	Turbines designed to be compliant and shed loads Optimize for performance Low solidity, optimum blade pitch One or two blades Higher rotor tip speed ratio
C	Turbines designed to manage loads mechanically and/or electrically Optimize for control Mechanical and electrical innovations (flapping or hinged blades, variable speed generators, etc.) Two or three blades Moderate rotor tip speed ratio

Modern grid-connected wind turbines usually follow the "C" approach, as it results in, better power quality, lower tip-speed ratios than approach "B", hence, lower visual disturbances and lower material requirements than in approach "A", as the structure does not need to withstand high wind loads, hence lower cost. Also, companies investigate combinations of the different approaches. However, the "C" approach currently dominates the commercial market. Each of the design approaches leaves a high degree of freedom regarding certain design details. For example, depending on the wind environment, different aerodynamic rotor diameters can be utilized. On high-wind speed sites, usually smaller rotor diameters are used with an aerodynamic profile that will reach the maximum efficiency between 14–16m/s. For low-wind sites, larger rotors are used but with an aerodynamic profile that will reach the maximum efficiency already between 12–14m/s. In both cases, the aim is to maximise the yearly energy harvest. In addition, wind turbine manufactures have to consider the overall cost, including the maintenance cost over the lifetime of the wind turbine.

Currently, three-bladed wind turbines dominate the market for grid-connected, horizontal-axis wind turbines. Two-bladed wind turbines, however, have the advantage that the tower top weight is lighter and, therefore, the whole supporting structure can be built lighter, with lower costs.

Three-bladed wind turbines have the advantage that the rotor moment of inertia is easier to understand and, therefore, often better to handle than the rotor moment of inertia of a two-bladed turbine. Furthermore, three-bladed wind turbines are often attributed "better" visual aesthetics and a lower noise level than two-bladed wind turbines. Both aspects are important considerations for wind turbine utilization in highly populated areas.

Wind turbines reach the highest efficiency at the designed wind speed, which is usually between 12m/s to 16m/s. At this wind speed, the power output reaches the rated capacity. Above this wind speed, the power output of the rotor must be limited to keep the power output close to the rated capacity and thereby reduce the driving forces on the individual rotor blade as well as the load on the whole wind turbine structure. Three options for the power output control are currently used:

Stall regulation: This principle requires a constant rotational speed, i.e., independent of the wind speed. A constant rotational speed can be achieved with a grid-connected induction generator. Due to the airfoil profile, the air stream conditions at the rotor blade change in a way that the air stream creates turbulence in high wind speed conditions, on the side of the rotor blade that is not facing the wind. This effect is known as stall effect. The effect results in a reduction of the aerodynamic forces and, subsequently, of the power output of the rotor. The stall effect is a complicated dynamic process. It is difficult to calculate the stall effect exactly for unsteady wind conditions. Therefore, the stall effect was for a long time considered to be difficult to use for large wind turbines.

However, due to the experience with smaller and medium-sized turbines, blade designers have learned to calculate the stall phenomenon more reliably. Today, even some manufacturers of megawatt turbines use stall-regulation, but the first prototypes of multi-megawatt wind turbines still avoid stall regulation. Figure 58 shows the output characteristics typical of a wind turbine using stall control.

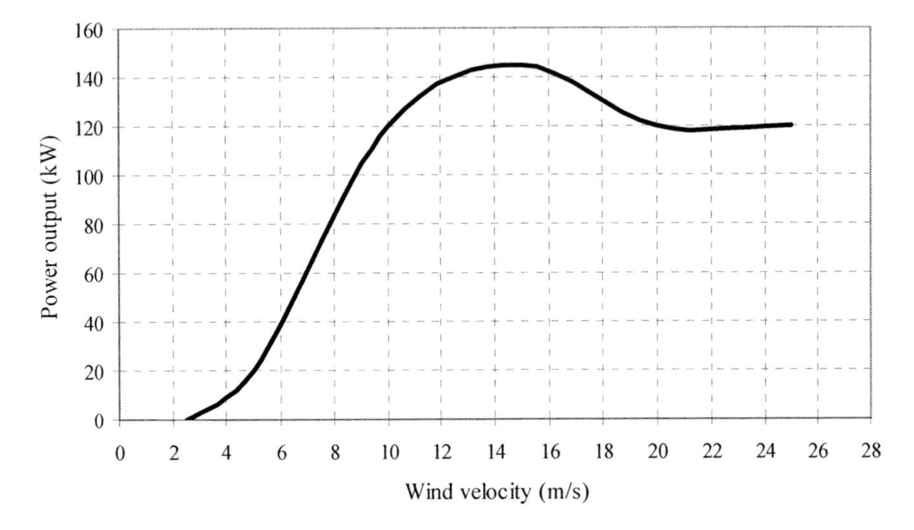

Figure 58. Typical power output chart of a turbine using stall control.

Pitch regulation: By pitching the rotor blades around their longitudinal axis, the relative wind conditions and, subsequently, the aerodynamic forces are affected in a way so that the power output of the rotor remains constant after rated power is reached. The pitching system in medium and large grid-connected wind turbines is usually based on a hydraulic system, controlled by a computer system. Some manufacturers also use electronically controlled electric motors for pitching the blades.

This control system must be able to adjust the pitch of the blades by a fraction of a degree at a time, corresponding to a change in the wind speed, in order to maintain a constant power output. The thrust of the rotor on the tower and foundation is substantially lower for pitch-controlled turbines than for stall-regulated turbines. In principle, this allows for a reduction of material and weight, in the primary structure. Pitch-controlled turbines achieve a better yield at low-wind sites than stall-controlled turbines, as the rotor blades can be constantly kept at optimum angle even at low wind speeds. Stall-controlled turbines have to be shut down once a certain wind speed is reached, whereas pitch-controlled turbines can gradually change to a spinning mode as the rotor operates in a no-load mode, i.e., it idles, at the maximum pitch angle. Figure 59 shows the output characteristics typical of a wind turbine using pitch control.

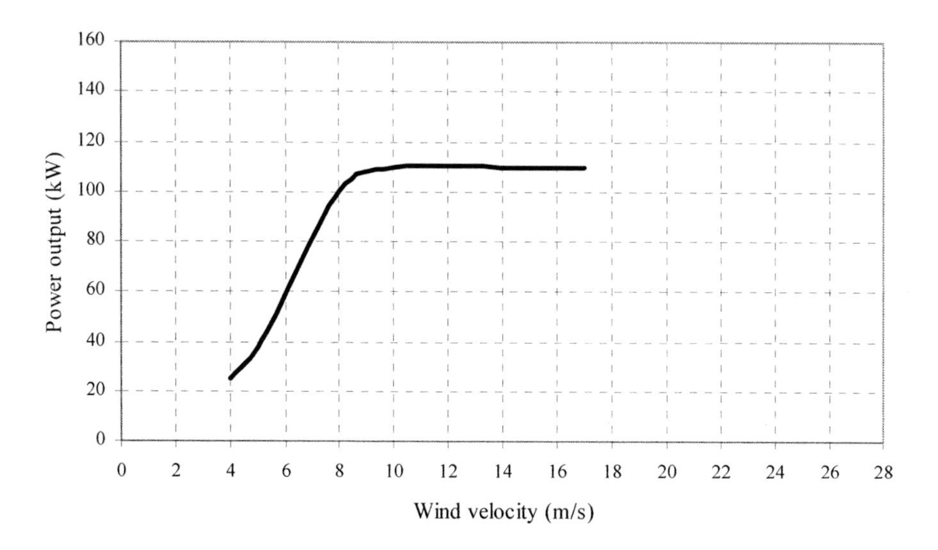

Figure 59. Typical output characteristics of a wind turbine using pitch control.

Active stall regulation: This regulation approach is a combination between pitch and stall. At low wind speeds, blades are pitched like in a pitch-controlled wind turbine, in order to achieve a higher efficiency and to guarantee a reasonably large torque to achieve a turning force. When the wind turbine reaches the rated capacity, the active stall-regulated turbine will pitch its blades in the opposite direction than a pitch-controlled machine does. This movement will increase the angle of attack of the rotor blades in order to make the blades go into a deeper stall. It is argued that active stall achieves a smoother limiting of power output, similar to that of pitch-controlled turbines without their difficult regulating characteristics.

It preserves, however, the advantage of pitch-controlled turbines to turn the blade into the low-load 'feathering position', hence thrust on the turbine structure is lower than on a stall-regulated turbine. If the wind speed reaches the cut-out wind speed (usually between 20m/s and 30m/s), the wind turbine shuts off and the entire rotor is turned out of the wind to protect the overall turbine structure. Because of this procedure, possible energy that could have been harvested will be lost. However, the total value of the lost energy over the lifetime of the wind turbine will usually be smaller than the investments that will be avoided by limiting the strength of the turbine to the cut-out speed.

The wind energy advantages can be summarized as:

- no fuel is used,
- no wastes or greenhouse gases are produced,
- the land beneath can usually still be used for farming,
- can supply energy to remote areas and
- maintenance requirements are minimal.

The wind energy disadvantages can be summarized as:

- the wind is not always predictable (e.g., some days have no wind),
- some people feel that covering the landscape with these towers is unsightly,
- can kill birds since migrating flocks tend to like strong winds,
- can affect nearby houses television reception,
- Can be noisy since wind generators have a reputation for making a constant, low, "swooshing" noise day and night and
- suitable areas for wind farms are often near the coast, where land is expensive.

Current research and development in the field of wind technologies include:

- the reduction of wind turbine weight,
- the reduction of noise,
- the development of methodologies for short term wind power forecasting,
- the utilization of low wind potential locations and
- the reduction in the kWh cost.

5.2. Biomass Energy

Biomass covers a wide range of products, by-products and waste streams from forestry and agriculture (including animal husbandry) as well as municipal and industrial waste streams. A definition adopted by legislation is: "...the biodegradable fraction of products, waste and residues from agriculture (including vegetal and animal substances), forestry and related industries, as well as the biodegradable fraction of industrial and municipal waste...". Biomass thus includes trees, arable crops, algae and other plants, agricultural and forest residues, effluents, sewage sludge, manures, industrial by-products and the organic fraction of municipal solid waste.

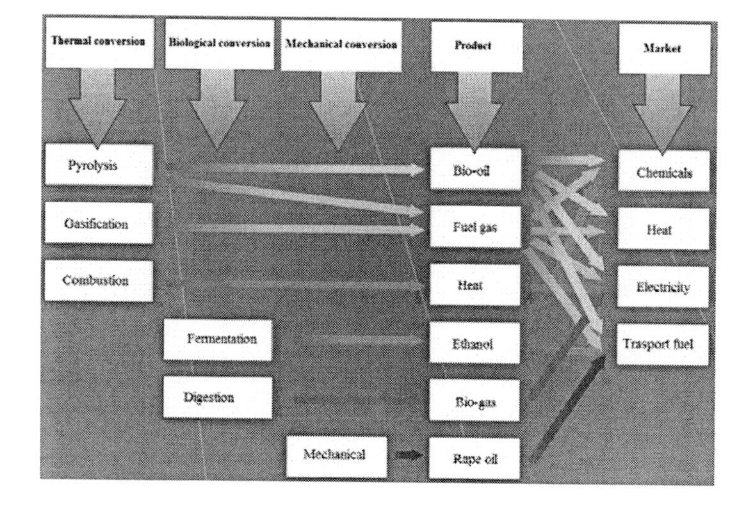

Figure 60. Biomass conversion pathways.

There are three ways to use biomass as shown in Figure 60. It can be burned to produce heat and electricity; changed to gas-like fuels, such as methane, hydrogen, and carbon monoxide; or changed to a liquid fuel. Liquid fuels, also called bio-fuels, include mainly two forms of alcohol: ethanol and methanol. The two most common bio-fuels are ethanol and bio-diesel.

The most commonly used bio-fuel is ethanol, which is produced from sugarcane, corn, and other grains. A blend of gasoline and ethanol is already used in cities with high air pollution. However, ethanol made from biomass is currently more expensive than gasoline on a gallon-for-gallon basis. Ethanol is mostly used as a fuel additive or oxygenate to enhance the octane and to cut down a vehicle's carbon monoxide and other smog-causing emissions. So, it is very important for scientists to find less expensive ways to produce ethanol from other biomass crops.

Bio-diesel can be used as a diesel additive to reduce vehicle emissions or in its pure form to fuel a vehicle. Concerns about the depletion of diesel fuel reserves and the pollution caused by continuously increasing energy demands make bio-diesel an attractive alternative motor fuel for compression ignition engines.

Heat is used to convert biomass into a fuel oil, which is then burned like petroleum to generate electricity. Biomass can also be burned directly to produce steam for electricity production or manufacturing processes. In industrialized countries, the main biomass processes utilized in the future are expected to be the direct combustion of residues and wastes for electricity generation. The future of biomass electricity generation lies in biomass integrated gasification/gas turbine technology, which offers high-energy conversion efficiencies. The electricity is produced by the direct combustion of biomass; advanced gasification and pyrolysis technologies are almost ready for commercial-scale use. Biomass power plants use technology that is very similar to that used in coal-fired power plants.

The biomass advantages can be summarized as:

- waste materials are used,
- the fuel, in some cases, tends to be cheap and
- less demand on the Earth's resources.

The biomass disadvantages can be summarized as:

- collecting the waste in sufficient quantities can be difficult,
- fuel is burned, so greenhouse gases are emitted and
- some waste materials are not available all year round.

Current research and development in the field of biomass includes:

- the improvement of the various technologies efficiency,
- the development of reliable and cost effective gasification systems,
- the development of new methods for cost effective production of clean bio-fuels for use in combustion engines and fuel cells and
- the reduction of costs (i.e., for bio-fuels the cost to be reduced at 36€c/kWh by the year 2020).

5.3. Geothermal Energy

The interior of the Earth is hot. The temperature in the Earth increases gradually with depth and reaches 4500°C at the center of the Earth. In general, the temperature rises one degree Celsius for every 36m depth. At some exceptional locations, hot rocks at a temperature of several hundred degrees are found at a depth of just a few kilometers. Hot springs, steam vents, and geysers bring some of the heat from the interior of the Earth to the surface.

These natural sources of geothermal energy are being exploited in a few places such as near Pisa in Italy, the Geysers in California, Cerro Prieto in Mexico, and Wairakei in New Zealand. The geothermal power plants operate with steam blowing out from wells drilled down to depths of as much as 2100m. For example the total power generated by the Geysers power plant in California is 700MWe, which makes it the largest of all the geothermal sources in operation.

Geothermal sources near populated areas can be used directly to provide heat and steam for homes and for industrial processes. In Iceland, nearly half the population lives in houses heated by geothermal sources; almost all the houses in Reykjavik are heated in this manner.

It may be possible to supplement the natural supply of geothermal steam by the artificial exploitation of the thermal energy of hot rocks. In some sites, rocks at a temperature of several hundred degrees are near enough to the surface of the Earth to be reached by drilling. If the rock can be fractured with high-pressure water or with explosives, then water can be circulated through it. The water would absorb the heat of the rock and make steam.

In general geothermal plants emit very little air pollution and have minimal impacts on the environment. Geothermal energy for electricity generation has been produced commercially since 1913, and for four decades on the scale of hundreds of MW both for electricity generation and direct use. The utilization has increased rapidly during the last three decades. In 2008, geothermal resources have been identified in over 80 countries, and there are quantified records of geothermal utilization in 58 countries in the world.

The direct application of geothermal energy can involve a wide variety of end uses, such as space heating and cooling, industry, greenhouses, fish farming, and health spas. It uses mostly existing technology and straightforward engineering. The technology, reliability, economics, and environmental acceptability of the direct use of geothermal energy have been demonstrated throughout the world. Electricity is produced with geothermal steam in 21 countries spread over all continents. Low-temperature geothermal energy is exploited in many countries to generate heat, with an estimated capacity of about 10000MWth.

Geothermal energy is clean, cheap and renewable, and can be utilized in various forms, such as space heating and domestic hot water supply, CO_2 and dry-ice production process, heat pumps, greenhouse heating, swimming and balneology (therapeutic baths), industrial processes, and electricity generation.

The main types of direct use are bathing, swimming and balneology (42%), space heating (35%), greenhouses (9%), fish farming (6%), and industry (6%).

The geothermal energy advantages can be summarized as:

- no fuel is used,
- no wastes or greenhouse gases are produced and
- less space is required.

The geothermal energy disadvantages can be summarized as:

- no many places around the world in which geothermal potential is available,
- sometimes a geothermal site may "run out of steam", perhaps for decades and
- hazardous gases and minerals may come up from underground, and can be difficult to safely dispose of.

5.4. HYDROPOWER

Hydropower (also called hydroelectric power) facilities in the world can generate enough power to supply many households with electricity. Water in rivers and streams can be captured and turned into hydropower. Hydropower is also inexpensive, and like many other renewable energy sources, it does not produce air pollution. Researchers are working on advanced turbine technologies that will not only help maximize the use of hydropower, but also minimize adverse environmental effects. A variety of mitigation techniques are in use now, and environmentally friendly turbines are under development.

Large-scale hydropower provides about one-quarter of the world's total electricity supply, virtually all of Norway's electricity, and more than 40% of the electricity used in developing countries. The technically usable world potential of large-scale hydro is estimated to be over 2200GW, of which only about 25% is currently exploited. There are two small-scale hydropower systems: microhydropower systems, with capacities below 100kW, and small hydropower systems, with a capacity between 101kW and 1MW. Large-scale hydropower supplies 20% of global electricity. In developing countries, considerable potential still exists, but large hydropower projects may face financial, environmental, and social constraints. Table 15 shows the development of electricity generation from hydropower in the world between 1995-2010.

For a hydroelectric power plant a dam is built to trap water, usually in a valley where there is an existing lake as indicated in Figure 61. Water is then allowed to flow through tunnels in the dam, to turn turbines and thus drive generators.

Table 15. World electricity production from hydropower

Hydro scale	Electricity production in 1995 (TWh/y)	Estimated electricity production in 2010 (TWh/y)
Large hydro	2265	3990
Small hydro	115	220
Total	2380	4210

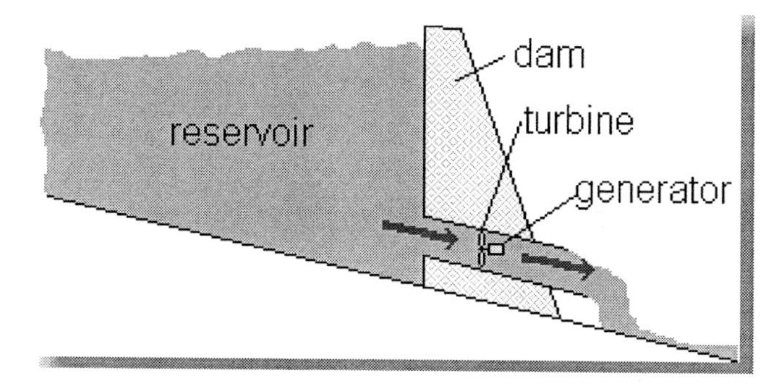

Figure 61. Hydropower principle.

Notice that the dam is much thicker at the bottom than at the top, because the pressure of the water increases with depth. Although there are many suitable sites around the world, hydro-electric dams are very expensive to build. However, once the station is built, the water comes free of charge and there is no waste or pollution.

The hydropower advantages can be summarized as:

- no fuel is used,
- no wastes or greenhouse gases are produced,
- much more reliable than wind or solar or wave power,
- water can be stored above the dam ready to cope with peaks in demand,
- hydro-electric power stations can increase to full power very quickly, unlike other power stations and
- electricity can be generated constantly.

The hydropower disadvantages can be summarized as:

- the dams are very expensive to build, however, many dams are also used for flood control or irrigation, so building costs can be shared,
- building a large dam will flood a very large area upstream, causing problems for animals that used to live there,
- finding a suitable site can be difficult - the impact on residents and the environment may be unacceptable and
- Water quality and quantity downstream can be affected, which can have an impact on plant life.

5.5. TIDAL ENERGY

Tides are caused by the gravitational attraction of the moon and the sun acting upon the oceans of the rotating earth. The relative motions of these bodies cause the surface of the oceans to be raised and lowered periodically, according to a number of interacting cycles. These include:

- a half day cycle, due to the rotation of the earth within the gravitational field of the moon,
- a 14 day cycle, resulting from the gravitational field of the moon combining with that of the sun to give alternating spring (maximum) and neap (minimum) tides,
- a half year cycle, due to the inclination of the moon's orbit to that of the earth, giving rise to maxima in the spring tides in March and September,
- other cycles, such as those over 19 years and 1600 years, arising from further complex gravitational interactions.

The range of a spring tide is commonly about twice that of a neap tide, whereas the longer period cycles impose smaller perturbations. In the open ocean, the maximum amplitude of the tides is about one metre. Tidal amplitudes are increased substantially towards the coast, particularly in estuaries. This is mainly caused by shelving of the sea bed and funnelling of the water by estuaries. In some cases the tidal range can be further amplified by reflection of the tidal wave by the coastline or resonance. This is a special effect that occurs in long, trumpet-shaped estuaries, when the length of the estuary is close to one quarter of the tidal wave length. For example, these effects combine to give a mean spring tidal range of over 11m in the Severn Estuary in UK. As a result of these various factors, the tidal range can vary substantially between different points on a coastline.

The amount of energy obtainable from a tidal energy scheme therefore varies with location and time. Output changes as the tide ebbs and floods each day; it can also vary by a factor of about four over a spring-neap cycle. Tidal energy is, however, highly predictable in both amount and timing.

The available energy is approximately proportional to the square of the tidal range. Extraction of energy from the tides is considered to be practical only at those sites where the energy is concentrated in the form of large tides and the geography provides suitable sites for tidal plant construction. Such sites are not commonplace but a considerable number have been identified in the UK, France, Eastern Canada, the Pacific coast of Russia, Korea, China, Mexico and Chile. Other sites have been identified along the Patagonian coast of Argentina, Western Australia and Western India.

Tidal barrages shown in Figure 62, work rather like a hydro-electric scheme, except that the dam is much bigger. A huge dam (called a "barrage") is built across a river estuary. When the tide goes in and out, the water flows through tunnels in the dam. The ebb and flow of the tides can be used to turn a turbine, or it can be used to push air through a pipe, which then turns a turbine.

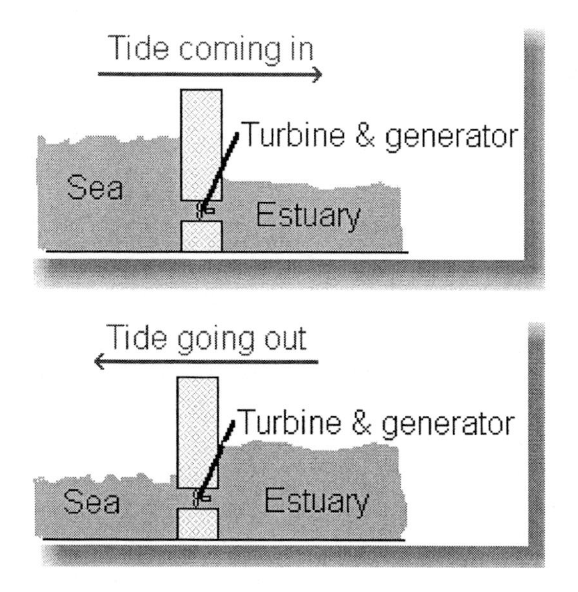

Figure 62. Tidal barrage principle.

The largest tidal power station in the world (and the only one in Europe) is the La Rance estuary in northern France, built in 1966. It consists of 24 bulb turbines with a capacity of 10MWe each.

A major drawback of tidal power stations is that they can only generate when the tide is flowing in or out - in other words, only for 10 hours each day. However, tides are totally predictable, so we can plan to have other power stations generating at those times when the tidal station is out of action.

Tidal energy projects based on barrages are capital-intensive with relatively high unit costs per installed kilowatt (greater than 2500US$/kW). The long construction period for the larger schemes and low load factors would result in high unit costs of energy, especially given the demands of private-sector investors. The economic performance of tidal energy barrages reflects the influence of site-specific conditions and the necessity for ship locks where access for navigation is required. As barrage construction is based upon conventional technology and site-specific conditions, it is unlikely that significant cost reductions could be achieved. Predicted unit costs of generation are therefore unlikely to change and currently remain uncompetitive with conventional fossil-fuel alternatives.

Some non-energy benefits would stem from the development of tidal energy schemes. However, they would yield a relatively minor monetary value in proportion to the total scheme cost. These benefits are difficult to quantify accurately and may not necessarily accrue to the barrage developer. Employment opportunities would be substantial at the height of construction, with the creation of some permanent long-term employment from associated regional economic development.

Tidal energy can also be exploited directly from marine currents induced by the combined lunar and solar gravitational forces responsible for tides. These forces cause semi-diurnal movement in water in shallow seas, particularly where coastal morphology creates natural constrictions, for example around headlands or between islands. This phenomenon produces strong currents, or tidal streams, which are prevalent around the British Isles and many other parts of the world where there are similar conditions. These currents are particularly prevalent where there is a time difference in tidal cycles between two sections of coastal sea. The flow is cyclical, increasing in velocity and then decreasing before switching to the opposite direction. The kinetic energy within these currents could be converted to electricity, by placing free-standing turbo-generating equipment in offshore areas as shown in Figure 63.

Figure 63. An under water tidal power station.

The benefits from utilizing marine currents are given below:

- marine currents have four times the energy density of a good wind site (diameter of water turbines less than half that of a wind turbine),
- the water velocities and therefore power outputs are completely predictable,
- water turbines will not need to be designed for extreme atmospheric fluctuations as required with wind turbines, (the design can be better cost-optimized),
- with increased conflicts over land use, water turbines offer a solution that will not occupy land and has minimal or zero visual impact and
- the technology is potentially modular and avoids the need for large civil engineering works.

Economic prospects for alternative forms of tidal energy remain uncertain, largely because there is little published data on the costs or performance of either marine current generators or bunded reservoir schemes. Until further information is made available it is not possible to make a rational judgement on their prospects. However, without detailed technical information (for investors) and rigorous appraisal of environmental effects no form of tidal energy is likely to be developed. Experience of other forms of renewable energy has highlighted the necessity for credible environmental assessment to ensure endorsement from regulatory authorities and potential objectors.

The tidal energy advantages can be summarized as:

- no fuel is used,
- no wastes or greenhouse gases are produced,
- tides are totally predictable, therefore, electricity is produced reliably and
- low maintenance costs.

The tidal energy disadvantages can be summarized as:

- high capital cost,
- change in the environment for many miles upstream and downstream,
- few suitable sites for tidal barrages and
- provides power for around 10 hours only each day, when the tide is actually moving in or out.

Current research and development in the field of tidal energy includes:

- the exploitation of marine currents induced by tides.

5.6. Wave Energy

Wave energy can be harnessed in coastal areas, close to the shore. Wave energy converters fixed to the shoreline are likely to be the first to be fully developed and deployed, but waves are typically 2–3 times more powerful in deep offshore waters than at the shoreline. Wave power technologies have been around for nearly thirty years. In fact the first patent for a wave energy device was filed in Paris in 1799, and by 1973, there were many patents for wave energy devices. Setbacks and a general lack of confidence have contributed to slow progress towards proven devices that would have a good probability of becoming commercial sources of electrical power.

Concerning the world resource for wave presented in Figure 64, the highest energy waves are concentrated off the western coasts in the 40°–60° latitude range north and south. The power in the wave fronts varies in these areas between 30kW/m and 70kW/m with peaks to 100kW/m in the Atlantic Southwest of

Ireland, the Southern Ocean and off Cape Horn. The capability to supply electricity from this resource is such that, if harnessed appropriately, 10% of the current level of world supply could be provided.

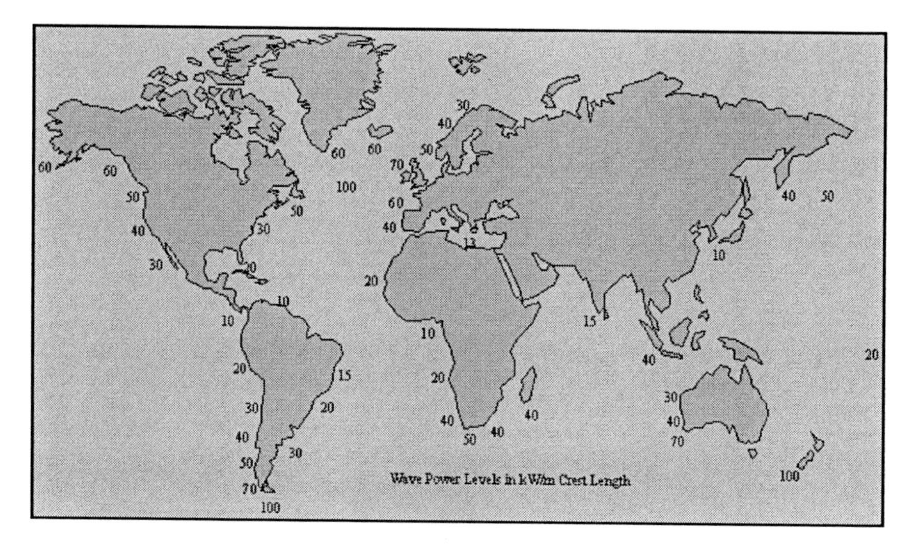

Figure 64. Global distribution of deep water wave power resources.

There are several methods of getting energy from waves, but one of the most effective works like a swimming pool wave machine in reverse. At a swimming pool, air is blown in and out of a chamber beside the pool, which makes the water outside bob up and down, causing waves. At a wave power station shown in Figure 65, the waves arriving cause the water in the chamber to rise and fall, which means that air is forced in and out of the hole in the top of the chamber. A turbine connected with a generator placed in this hole, is turned by the air rushing in and out producing electricity. A problem with this design is that the rushing air can be very noisy, unless a silencer is fitted to the turbine. The noise is not a huge problem anyway, as the waves make quite a bit of noise themselves.

A new approach which is under development known as pelamis (named after a sea-snake), consists of a series of cylindrical segments connected by hinged joints as indicated in Figure 66.

As waves run down the length of the device and actuate the joints, hydraulic cylinders incorporated in the joints pump oil to drive a hydraulic motor via an energy-smoothing system. Electricity generated in each joint is transmitted to shore by a common sub-sea cable. The slack-moored device will be around 130m long and 3.5m in diameter. The pelamis is intended for general deployment offshore and is designed to use technology already available in the offshore industry.

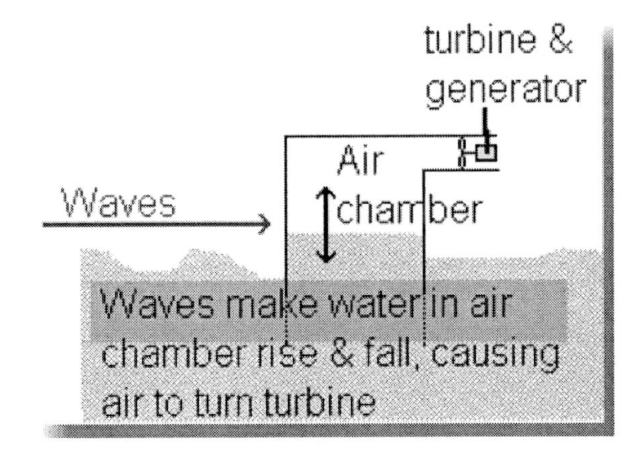

Figure 65. Wave power station principle.

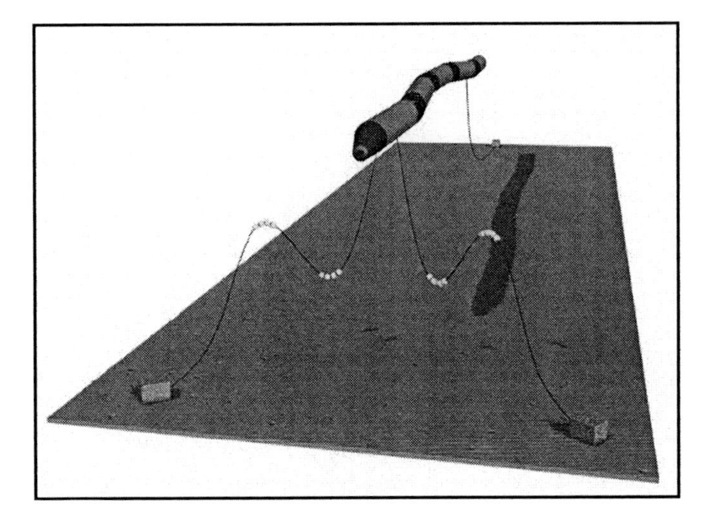

Figure 66. 3D view of the "Pelamis" concept.

The full-scale version has a continuously rated power output of 0.75MW. Currently a one-seventh-scale prototype is being prepared for deployment.

The wave energy advantages can be summarized as:

- no fuel is used,
- no wastes or greenhouse gases are produced and
- low maintenance costs.

The wave energy disadvantages can be summarized as:

- depends on the waves - sometimes many loads of energy available, sometimes nothing,
- a suitable site is required, where waves are consistently strong,
- some designs are noisy and
- must be able to withstand very rough weather.

Current research and development in the field of wave energy includes:

- the development of offshore wave energy collection concept, using a floating tube called "pelamis".

Chapter 6

6. DISTRIBUTED GENERATION

DG technologies have been available for many years. They may have been known by different names such as generators, back-up generators, or on-site power systems, and certain DG technologies are not new (e.g., internal combustion engines and gas turbines). On the other hand, due to the changes in the utility industry, several new technologies are being developed or advanced toward commercialization (e.g., fuel cells and photovoltaics).

In the past few years, DG technologies have made a growing number of excited claims that small generators will revolutionize the electricity generation sector and have an enormous environmental payoff. A future is envisioned in which DG technologies are as ubiquitous as boilers. Homeowners and businesses would buy these small generators and have them installed just as they would any other appliance. In these visions, DG technologies become so common that they enhance electric reliability to near perfection.

6.1. DG DEFINITION

In the early days of electricity generation, DG was the rule, not the exception. The first power plants only supplied electricity to customers in the close neighborhood of the generation plant. The first grids were DC based, and therefore, the supply voltage was limited, as was the distance that could be used between generator and consumer. Balancing demand and supply was partially done using local storage, i.e., batteries, which could be directly coupled to the DC grid. Subsequent technology developments driven by economies of scale resulted in the development of large centralized grids connecting up entire regions and countries. The design and operating philosophies of power systems have emerged with a focus on centralized generation. During the last decade, there has been renewed interest in DG.

Although, DG is a new approach in the electricity industry there is no generally accepted definition, but many definitions exist. A short survey of the literature shows that there is no consensus on DG definition. Some countries define DG on the basis of the voltage level, whereas others start from the principle that DG is connected to circuits from which consumer loads are supplied directly. Other countries, define DG as having some basic characteristic (e.g., using renewables, cogeneration, etc.). Some definitions allow for the inclusion of larger-scale cogeneration units or large wind farms connected to the transmission grid, others put the focus on small-scale generation units connected to the distribution grid. All these definitions suggest that at least the small-scale generation units connected to the distribution grid are to be considered as part of DG. Moreover, generation units installed close to the load or at the customer side of the meter are also commonly identified as DG. This latter criterion partially overlaps with the first, as most of the generation units on customer sites are also connected to the distribution grid. However, it also includes somewhat larger generation units, installed on customer sites, but connected to the transmission grid. In regards to the capacity of DG technologies different scenarios can be found ranging from a few kW to 100MWe. In order to obtain a unified definition of DG technologies the following DG issues were examined:

(a) purpose,
(b) location,
(c) capacity,
(d) power delivery area,
(e) technology,
(f) environmental impact,
(g) mode of operation,
(h) ownership and
(i) level of penetration.

A general DG definition which is now widely accepted is as follows:

Distributed Generation is an electric power source connected directly to the distribution network or on the customer site of the meter.

The distinction between distribution and transmission networks is based on the legal definition, which is usually part of the electricity market regulation of each country. Anything that is not defined as transmission network in the legislation can be regarded as distribution network. The above definition of DG does not define the rating of the generation source, as the maximum rating depends on the local distribution network conditions, e.g., voltage level.

It is, however, useful to introduce categories of different ratings of distributed generation. The following categories are suggested:

(a) Micro DG, 1We – 5kWe,
(b) Small DG, 5kWe – 5MWe,
(c) Medium DG, 5MWe – 50MWe,
(d) Large DG, 50MWe – 300MWe.

Also, the definition of DG does not define the technologies, as the technologies that can be used vary widely. However, a categorization of different technology groups of DG seems possible:

(a) Non-renewable DG,
(b) Renewable DG.

The different types and technologies that can be used for DG applications are illustrated in Figure 67.

6.2. The Fuel Cell Technology

A fuel cell is an energy conversion device that generates electricity and heat by electrochemically combining a gaseous fuel (hydrogen) and an oxidant gas (oxygen from the air) through electrodes and across an ion conducting electrolyte. During this process, water is formed at the exhaust. The fuel cell does not run down or require any recharging, unlike a battery it will produce energy as long as fuel is supplied. The principle characteristic of a fuel cell is its ability to convert chemical energy directly to electrical energy. This gives much higher conversion efficiencies than any conventional thermo–mechanical system. Therefore, fuel cells extract more electricity from the same amount of fuel, to operate without combustion so they are virtually pollution free and have quieter operation since there are no moving parts.

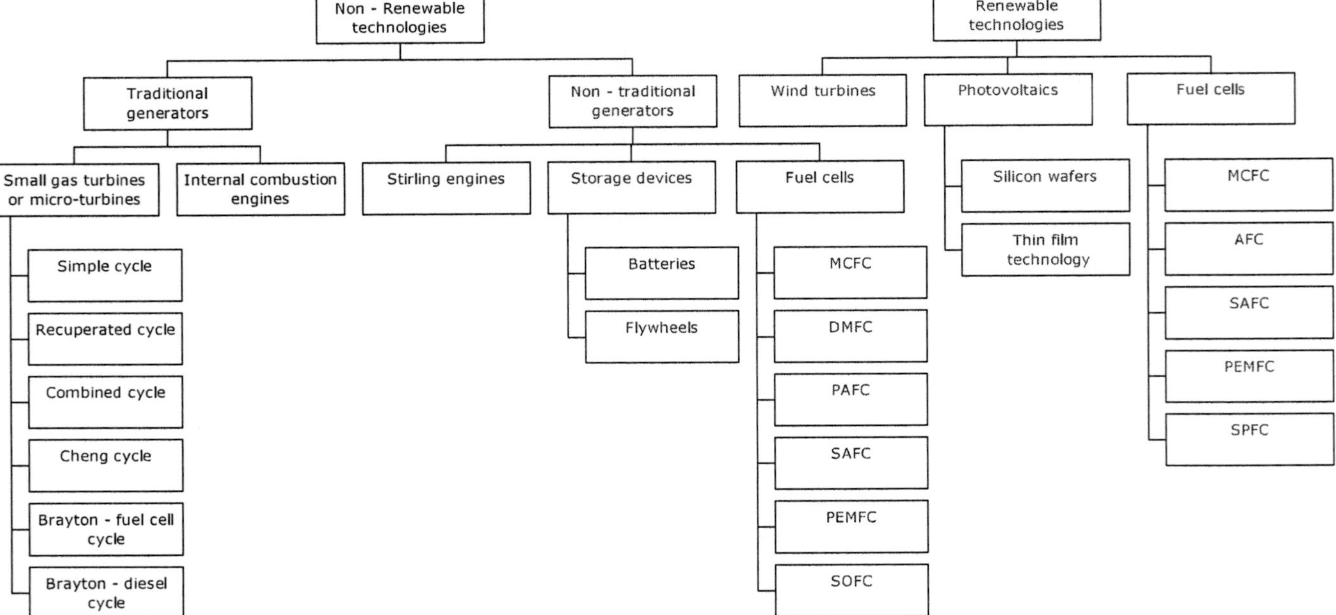

* DER refers to distributed energy resources which a wider definition of DG.

Figure 67. DER technologies for power generation.

The fuel cell uses oxygen and hydrogen to produce electricity. The oxygen comes from the air (present at around 20%) unlike the hydrogen, which is difficult to store and distribute, and this is the reason for which hydrocarbon or alcohol fuels, readily available, are used. A reformer is, therefore, needed to turn these products into hydrogen, which is then fed to the fuel cell. Some of the fuel cells have problems with electrolyte management (liquid electrolytes, for example, which are corrosive and difficult to handle), others use expensive material such as platinum as in the Proton Exchange Membrane Fuel Cells (PEMFC), need hydration of their electrolyte material or have a high operating temperature which is the case of the Solid Oxide Fuel Cells (SOFC) and Molten Carbonate Fuel Cells (MCFC).

Fuel cells provide highly efficient, pollution free power generation. Their performance has been confirmed by successful operation power generation systems. Electrical-generation efficiencies of 70% are possible along with a heat recovery possibility, e.g., the Brayton – fuel cell cycle (for details see section 2.9 The Brayton – fuel cell cycle on page 55). It is expected that in the future, technology will open up new possibilities and fuel cell based power systems will be ideal power-generation systems, being reliable, clean, quiet, environmentally friendly, and fuel conserving.

A fuel cell consists of two electrodes sandwiched around an electrolyte. Hydrogen fuel is fed into the anode of the fuel cell and oxygen, from the air, enters the cell through the cathode. The hydrogen, under the action of the catalyst, splits into protons (hydrogen ions) and electrons, which take different paths towards the cathode. The proton passes through the electrolyte and the electron create a separate current that can be used before reaching the cathode, to be reunited with the hydrogen and oxygen to form a pure water molecule and heat as shown in Figure 68.

In more detail, the fuel cell is mainly composed of two electrodes, the anode and the cathode, the catalyst, and an electrolyte. The main function of the electrode is to bring about reaction between the reactant (fuel or oxygen) and the electrolyte without itself being consumed or corroded. It must, also, bring into contact the three phases, i.e., the gaseous fuel, the liquid or solid electrolyte and the electrode itself. The anode, used as the negative post of the fuel cell, disperses the hydrogen gas equally over the whole surface of the catalyst and conducts the electrons that are freed from hydrogen molecule, to be used as a useful power in an external circuit. The cathode, the positive post of the fuel cell, distributes the oxygen fed to it onto the surface of the catalyst and conducts the electrons back from the external circuit where they can recombine with hydrogen ions, passed across the electrolyte, and oxygen to form water.

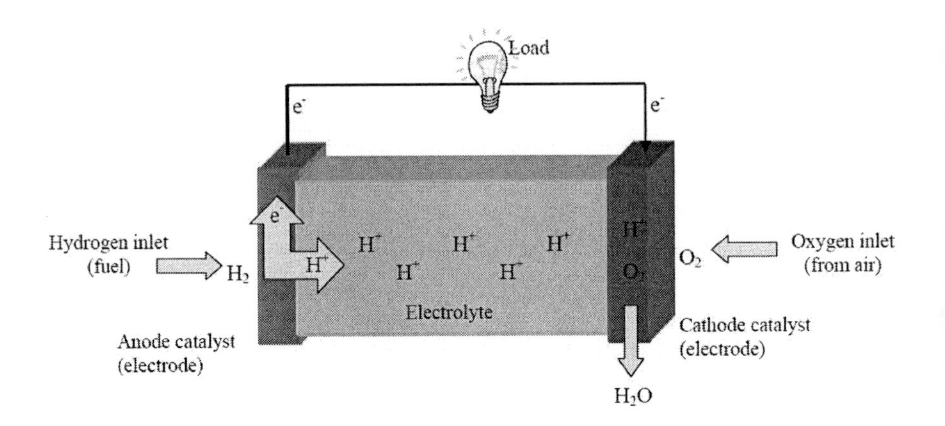

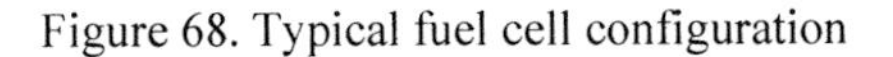
Figure 68. Typical fuel cell configuration

The catalyst is a special material that is used in order to facilitate the reaction of oxygen and hydrogen. This can be a platinum coating as in PEMFC or nickel and oxide for the SOFC. The nature of the electrolyte, liquid or solid, determines the operating temperature of the fuel cell. It is used to prevent the two electrodes, by blocking the electrons, to come into electronic contact. It, also, allows the flow of charged ions from one electrode to the other. It can either be an oxygen ion conductor or a hydrogen ion (proton) conductor, the major difference between the two types is the side in the fuel cell in which the water is produced; the oxidant side in proton conductor fuel cells and the fuel side in oxygen-ion-conductor ones.

The fuel cells are sorted by their operating temperature and their classification is generally done according to the nature of the electrolyte used. There are several types of fuel cell technologies being developed for different applications, each using a different chemistry, as summarized in Table 16.

There are, also, other types of fuel cells which are less employed but may later find a specific application, for example, the air-depolarised cells, sodium amalgam cells, biochemical fuel cells, inorganic redox cells, regenerative cells, alkali metal–halogen cells, etc. Practical fuel cells can be combined to form a fuel cells' stack. The cells are connected in electrical series to build a desired output voltage. An interconnect component connects the anode of one cell to the cathode of the next cell in the stack. A fuel cells' stack can be configured in series, parallel, series–parallel or as single units, depending upon the type of applications. The number of fuel cells in a stack determines the total voltage, and the surface of each cell gives the total current.

Present material science has made the fuel cells a reality in some specialized applications. By far the greatest research interest throughout the world has focused on PEMFC and SOFC stacks. PEMFCs are a well advanced type of fuel cell that are suitable for cars and mass transportation if they can be made cost competitive. Their efficiency is expected to reach around 50%, which is better than any internal combustion engine.

Table 16. Technical characteristics of different types of fuel cells

Type	Electrolyte	Efficiency (%)	Operating temper. (P^{oP}C)	Fuel
Alkaline (AFC)	Potassium hydroxide (KOH)	N/A	50-200	Pure hydrogen or hydrazine
Direct methanol (DMFC)	Polymer	N/A	60-200	Liquid methanol
Phosphoric acid (PAFC)	Phosphoric acid	38	160-210	Hydrogen from hydrocarbons and alcohol
Sulphuric acid (SAFC)	Sulphuric acid	N/A	80-90	Alcohol or impure hydrogen
Proton-exchange membrane (PEMFC)	Polymer, proton exchange membrane	34	50-80	Less pure hydrogen from hydrocarbons or methanol
Molten carbonate (MCFC)	Molten salt such as nitrate, sulphate, carbonates, etc.	48	630-650	Hydrogen, carbon monoxide natural gas, propane, marine diesel
Solid oxide (SOFC)	Stabilized zirconia and doped perovskite	47	600-1000	Natural gas or propane
Solid polymer (SPFC)	Solid sulphonated polystyrene	N/A	90	Hydrogen

As for the future development of SOFCs, having efficiency around 70% with a heat conversion possibility, it is mainly concerned with reducing their operating temperature since expensive high temperature alloys are used to house the fuel cell. The reduction in the temperature will, therefore, allow the use of cheaper structural components such as stainless steel. A lower temperature will also ensure a greater overall system efficiency and a reduction in the thermal stresses in the active ceramic structures, leading to a longer expected lifetime of the system, and making possible the use of cheaper interconnect materials such as ferritic steels, without protective coatings.

Chapter 7

7. STORAGE TECHNOLOGIES

In today's world, the need for more energy seems to be ever-increasing. Both households and industries require large amounts of power. At the same time the existing means of energy production face new problems. International treaties aim to limit the levels of pollution, global warming prompts action to reduce the output of carbon dioxide and several countries have decided to decommission old nuclear power plants and not build new ones. In addition, the unprecedented global increase in energy demand has meant that the price of conventional energy sources has risen dramatically and that the dependence of national economies on a continuous and undistorted supply of such sources has become critical.

Such development brings about the need to replace old energy production methods with new ones. While several are in development, including the promising nuclear fission power, other production methods are already in commercial use. The penetration of renewable energy sources and of other forms of potential distributed generation sources is increasing worldwide. These types of energy sources often rely on the weather or climate to work effectively, and include such methods as wind power, solar power and hydroelectricity in its many forms.

These new sources of energy have some indisputable advantages over the older methods. At the same time, they present new challenges. The output of the traditional methods is easy to adjust according to the power requirements. The new energy sources are based more directly on harnessing the power of the nature and as such their peak power outputs may not match the power requirements. They may exhibit large fluctuations in power output in monthly or even annual cycles. Similarly, the demand can vary monthly or annually. Therefore, in order for these new sources to become completely reliable as primary sources of energy, energy storage is a crucial factor. Essentially, energy from these sources must be stored when excess is produced and then released, when production levels are less than the required demand. Energy storage technologies form therefore an integral and indispensable part of a reliable and effective renewable and distributed generation unit.

There are other reasons why it is necessary to store large amounts of energy. Depending on how storage is distributed, it may also help the network withstand peaks in demand. Storing energy allows transmission and distribution to operate at full capacity, decreasing the demand for newer or upgraded lines and increasing plant efficiencies. Storing energy for shorter periods may be useful for smoothing out small peaks and sags in voltage.

There is clearly a need for energy storage, specifically energy storage in a larger scale than before. Traditional energy storage methods, such as the electrochemical cell, are not necessarily applicable to larger-scale systems, and their efficiency may be suboptimal. Meanwhile, a number of new and promising methods are in development. Some of these are based on old concepts applied to modern energy storage, others are completely new ideas. Some are more mature than others, but most can be further improved.

7.1. Flywheel Storage Technologies

A flywheel is a mass rotating about an axis, which can store energy mechanically in the form of kinetic energy. Energy is required to accelerate the flywheel so it is rotating. This is usually achieved by an electric motor when being used in an electrical system. Once it is rotating, it is in effect a mechanical battery that has a certain amount of energy that can be stored depending on its rotational velocity and its moment of inertia. The faster a flywheel rotates the more energy it stores. This stored energy can be retrieved by slowing down the flywheel via a decelerating torque and returning the kinetic energy to the electrical motor, which is used as a generator.

Apart from the rotating flywheel, the other main components of a flywheel storage system are the rotor bearings and the power interface as illustrated in Figure 69. The flywheel can be either low speed, with operating speeds up to 6000rpm, or high-speed with operating speeds up to 50000rpm. Low speed flywheels are usually made of steel rotors and conventional bearings. Typical specific energy achieved is around 5Wh/kg. High-speed flywheels use advanced composite materials for the rotor with ultra-low friction bearing assemblies. These light-weight and high-strength composite rotors can achieve specific energy of 100Wh/kg. Also, such flywheels come up to speed in a matter of minutes, rather than the hours needed to recharge a battery. The container for high-speed flywheels is evacuated or helium filled to reduce aerodynamic losses and rotor stresses.

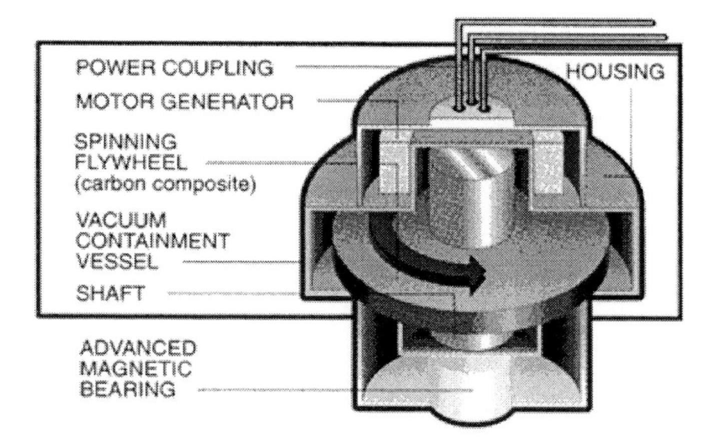

Figure 69. Main components of a flywheel storage system.

The power interface includes the motor/generator, a variable-speed power electronics converter and a power controller. The motor/generator is usually a high speed permanent magnet machine integrated with the rotor functioning as an integrated synchronous generator. The converter is usually a pulse width modulated bi-directional converter which can be single-stage (a.c to d.c) or double-stage (a.c to d.c to a.c) depending on the application requirements. Finally, a power controller is required to control power system variables.

The main advantages of flywheel storage systems are the high charge and discharge rates for many cycles. Indeed, the high cycling capability of flywheels is one of their key features and is not dependent on the charge or discharge rate. Full-cycle lifetimes range from 10^5 up to 10^7. In fact, the limiting factor in some applications is more likely to be the flywheel lifetime which is quoted as typically 20 years. Also, typical state-of-the-art composite rotors have high specific energies, up to 100Wh/kg, with high specific power. Their energy efficiency is typically around 90% at rated power. The main disadvantages of flywheels are the high cost and the relatively high standing losses. Self-discharge rates for complete flywheel systems are high, with minimum rate of 20% of the stored capacity per hour. These high rates have the effect of deteriorating energy efficiency when cycling is not continuous, for example when energy is stored for a period between charge and discharge. Such high discharge rates reinforce the notion that flywheels are not an adequate device for long-term energy storage but only to provide reliable standby power.

Such applications can be the integration of a flywheel energy storage system with a renewable energy source power plant system. The amount of power produced by renewable energy sources such as photovoltaic cells and wind turbines varies significantly on an hourly, daily and seasonal basis due to the variation in the availability of the sun, wind and other renewable resources. Even when conventional technologies are generating electricity at a constant rate, there are demand fluctuations throughout the day. This mismatch of load to electrical supply means that power is not always available when it is required and on other occasions, there is excess power. Flywheel technologies can be used to provide power when there is insufficient power being generated, and to store excess production. Another important application for flywheel technologies is for power conditioning and for providing power when there are durations of total power loss as a result of electricity grid failure.

7.2. Battery Storage Technologies

Storage batteries are rechargeable electrochemical systems used to store energy. They deliver, in the form of electric energy, the chemical energy generated by electrochemical reactions. These reactions are set in train inside a basic cell, between two electrodes plunged into an electrolyte, when a load is connected to the cell's terminals. The reaction involves the transfer of electrons from one electrode to the other through an external electric circuit/load.

A battery consists of single or multiple cells, connected in series or in parallel or both depending on the desired output voltage and capacity. Each cell, shown in Figure 70, consists of:

- The anode or negative electrode which provides electrons to the load and is oxidised during the electrochemical reaction
- The cathode or positive electrode which accepts electrons and is reduced during the reaction
- The electrolyte which provides the medium for transfer of electrons between the anode and the cathode
- The separators between positive and negative electrodes for electrical insulation

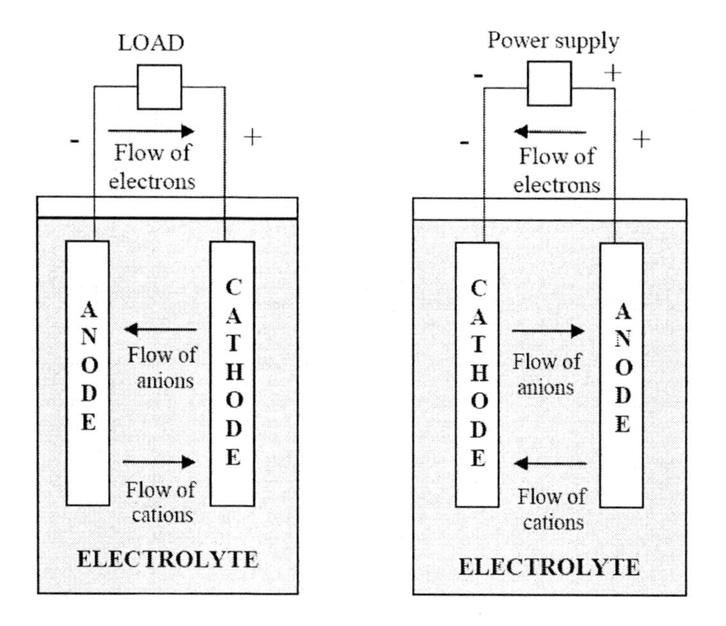

Figure 70. Chargeable cell/battery diagram.

There are three main types of conventional storage batteries that are used extensively today: the lead-acid batteries, the nickel-based batteries and the lithium-based batteries. Lead-acid batteries are the oldest type of rechargeable batteries and are based on chemical reactions involving lead dioxide (which forms the cathode electrode), lead (which forms the anode electrode) and sulphuric acid which acts as the electrolyte. The rated voltage of a lead-acid cell is 2V and typical energy density is around 30Wh/kg with power density around 180W/kg. Lead-acid batteries have high energy efficiencies (between 85-90%), are easy to install and require relatively low level of maintenance and low investment cost. In addition, the self-discharge rates for this type of batteries are very low, around 2% of rated capacity per month (at 25°C) which makes them ideal for long-term storage applications. However, the limiting factors for these batteries are the relatively low cycle life and battery operational lifetime. Typical lifetimes of lead-acid batteries are between 1200-1800 charge/discharge cycles or 5-15 years of operation. The cycle life is negatively affected by the depth of discharge and temperature. Attempts to fully discharge the battery can be particularly damaging to the electrodes, thus reducing lifetime. Regarding temperature levels, although high temperatures (up to 45°C which is the upper limit for battery operation) may improve battery performance in terms of higher capacity, they can also reduce total battery lifetime as well as the battery energy efficiency.

The nickel-based batteries are mainly the nickel-cadmium (NiCd), the nickel-metal hydride (NiMH) and the nickel-zinc (NiZn) batteries. All three types use the same material for the positive electrode and the electrolyte which is nickel hydroxide and an aqueous solution of potassium hydroxide with some lithium hydroxide respectively. As for the negative electrode, the NiCd type uses cadmium hydroxide, the NiMH uses a metal alloy and the NiZn uses zinc hydroxide. The rated voltage for the alkaline batteries is 1.2V (1.65V for the NiZn type) and typical maximum energy densities are higher than for the lead-acid batteries. Typically, values are 50Wh/kg for the NiCd, 80Wh/kg for the NiMH and 60Wh/kg for the NiZn. Typical operational life and cycle life of NiCd batteries is also superior to that of the lead-acid batteries. At deep discharge levels, typical lifetimes for the NiCd batteries range from 1500 cycles for the pocket plate vented type to 3000 cycles for the sinter vented type. The NiMH and NiZn have similar or lower values to those of the lead-acid batteries.

Despite the above advantages of the NiCd batteries over the lead-acid batteries, NiCd and the rest of the nickel-based batteries have several disadvantages compared to the lead-acid batteries in terms of industrial use or for use in supporting renewable energy power systems. Generally, the NiCd battery is the only one of the three types of nickel-based batteries that is commercially used for industrial UPS applications such as in large energy storage for renewable energy systems. However, the NiCd battery may cost up to 10 times more than the lead-acid battery. On top of that, the energy efficiencies for the nickel batteries are lower than for the lead-acid batteries. The NiMH batteries have energy efficiencies between 65-70% while the NiZn have 80% efficiency. The energy efficiency of the NiCd batteries varies depending on the type of technology used during manufacture. For the vented type, the pocket plate have 60%, the sinter/PBE plate have 73%, the fibre plate 83% and the sinter plate have 73% energy efficiency. Finally, the sealed cylindrical type of NiCd batteries has 65% energy efficiency. Another dimension where the NiCd batteries are inferior to the lead-acid batteries is the self-discharge rate. Self discharge rates for an advanced NiCd battery are much higher than those for a lead-acid battery since they can reach more than 10% of rated capacity per month.

The third major type of battery storage technology is the lithium-based battery storage system. This technology has not yet been used for energy storage in the context of an uninterrupted power supply (UPS) system although such applications are being developed. Currently, lithium battery technology is typically used in mobile or laptop systems and in the near future it is envisaged to be used in hybrid or electric vehicles. Lithium technology batteries consist of two main types: lithium-ion and lithium polymer cells. Their advantage over the NiCd and lead-acid batteries is their higher energy density and energy efficiency, their lower self-discharge rate and extremely low maintenance required. Lithium-ion cells, with nominal voltage around 3.7V, have energy densities ranging from 80 to 150Wh/kg while for lithium-polymer cells it ranges from 100 to 150Wh/kg. Energy efficiencies range from 90 to 100% for both these technologies. Power density for lithium-ion cells ranges from 500 to 2000W/kg while for lithium-polymer it ranges from 50 to 250W/kg.

For lithium-ion batteries, self-discharge rate is very low at maximum 5% per month and battery lifetime can reach more than 1500 cycles. However, the lifetime of a lithium-ion battery is temperature dependent, with aging taking its toll much faster at high temperatures, and can be severely shortened due to deep discharges. This makes lithium-ion batteries unsuitable for use in back-up applications where they may become completely discharged. In addition, lithium-ion batteries are fragile and require a protection circuit to maintain safe operation. Built into each battery pack, the protection circuit limits the peak voltage of each cell during charge and prevents the cell voltage from dropping too low on discharge. In addition, the cell temperature is monitored to prevent temperature extremes. The maximum charge and discharge current on most packs are also limited. These precautions are necessary in order to eliminate the possibility of metallic lithium plating occurring due to overcharge.

Lithium-polymer battery lifetime can only reach about 600 cycles. Regarding its self-discharge, this is much dependent on temperature but it has been reported to be around 5% per month. Compared to the lithium-ion battery, the lithium polymer battery operational specifications dictate a much narrower temperature range, avoiding lower temperatures. However, lithium-polymer batteries are lighter, and safer with minimum self inflammability.

Currently research into lithium-based batteries is mainly concerned with cost reduction by use of cheaper materials, lifetime increase and reduction of high flammability, especially in the case of lithium-ion batteries. Cost is currently estimated to be between 900 and 1300$/kWh.

Apart from the three main types of batteries described above, a few additional types also exist albeit with low penetration in the market. These are the sodium sulphur (NaS) battery, the Redox flow storage system and the metal-air battery.

The NaS battery consists of liquid (molten) sulphur at the positive electrode and liquid (molten) sodium at the negative electrode as active materials separated by a solid beta alumina ceramic electrolyte as shown in Figure 71. The electrolyte allows only the positive sodium ions to go through it and combine with the sulphur to form sodium polysulfides. During discharge, sodium gives off electrons, while positive Na^+ ions flow through the electrolyte and migrate to the sulphur container. The electrons flow in the external circuit of the battery producing about 2V and then through the electric load to the sulphur container. Here the electron reacts with the sulphur to form S^- cations, which then forms sodium polysulfide after reacting with sodium ions. As the cell discharges, the sodium level drops. This process is reversible as charging causes sodium polysulfides to release the positive sodium ions back through the electrolyte to recombine as elemental sodium.

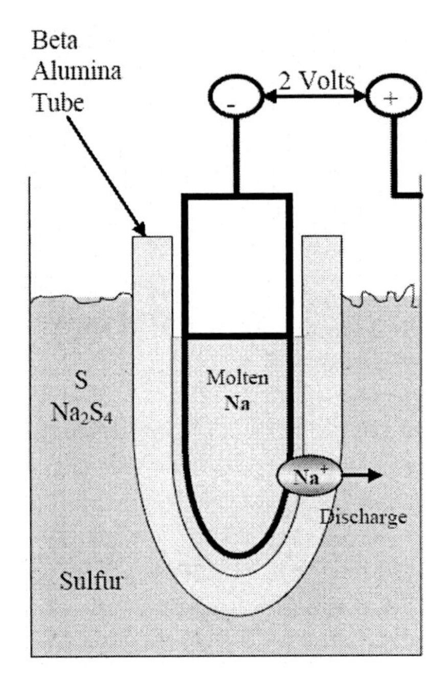

Figure 71. Basic 2V NaS cell/battery operation.

Once running, the heat produced by charging and discharging cycles is enough to maintain operating temperatures and no external heat source is required to maintain this process. Heat produced is typically about 300 to 350°C.

NaS batteries are highly energy efficient (89-92%) and are made from inexpensive and non-toxic materials. However, the high operating temperatures and the highly corrosive nature of sodium make them suitable only for large-scale stationary applications. NaS batteries are currently used in electricity grid related applications such as peak shaving and improving power quality.

Finally, the two other emerging battery storage technologies that are currently not yet used on a commercial basis are the Metal-air energy storage system and the Redox flow storage system. Both these technologies are under continuous research and technological development so as to become commercialized. Redox technology offers significant advantages such as no self-discharge and no degradation for deep discharge but it still faces technical development issues and also requires high investment cost. Metal-air technology offers high energy density (compared to lead-acid batteries), and long shelf life while promising reasonable cost levels. However, tests have shown that the metal-air batteries suffer from limited operating temperature range and a number of other technical issues not least of which is the difficulty in developing efficient, practical fuel management systems and cheap and reliable bifunctional electrodes.

7.3. Supercapacitor Storage Technologies

Supercapacitors (or ultracapacitors) are very high surface areas activated capacitors that use a molecule-thin layer of electrolyte as the dielectric to separate charge. The supercapacitor resembles a regular capacitor except that it offers very high capacitance in a small package. Supercapacitors rely on the separation of charge at an electric interface that is measured in fractions of a nanometer, compared with micrometers for most polymer film capacitors. Energy storage is by means of static charge rather than of an electro-chemical process inherent to the battery.

Depending on the material technology used for the manufacture of the electrodes, supercapacitors can be categorized into electrochemical double layer supercapacitors (ECDL) and pseudo-capacitors. Hybrid capacitors are also a new category of supercapacitors (see Figure 72 for supercapacitor taxonomy). ECDL supercapacitors are currently the least costly to manufacture and are the most common type of supercapacitor.

The main components of an ECDL supercapacitor can be observed in Figure 73. The ECDL supercapacitors have a double-layer construction consisting of carbon-based electrodes immersed in a liquid electrolyte (which also contains the separator). As electrode material, porous active carbon is usually used. Recent technological advancements have allowed carbon aerogels and carbon nanotubes to also be employed as electrode material. The electrolyte is either organic or aqueous. The organic electrolytes use usually acetonitrile and allow nominal voltage of up to 3 volts.

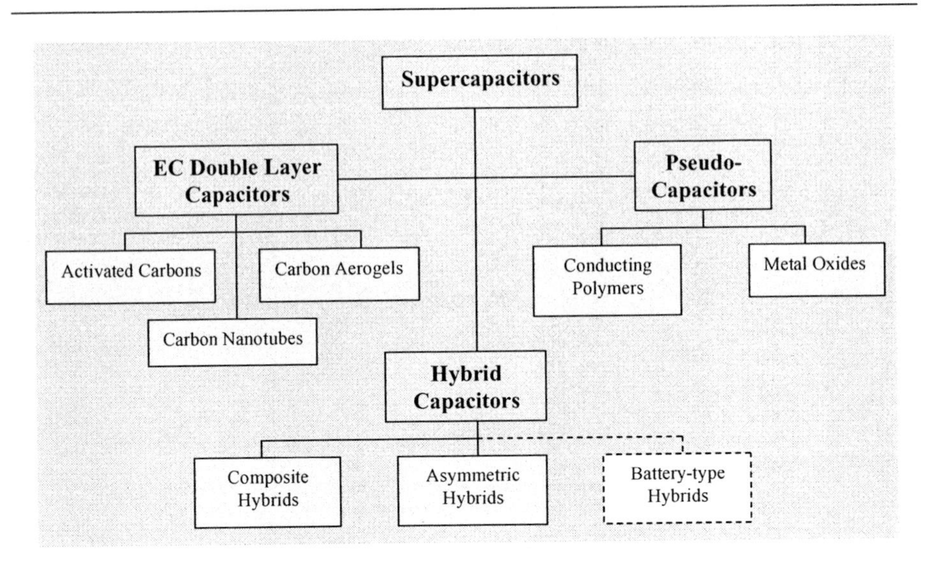

Figure 72. Taxonomy of supercapacitors.

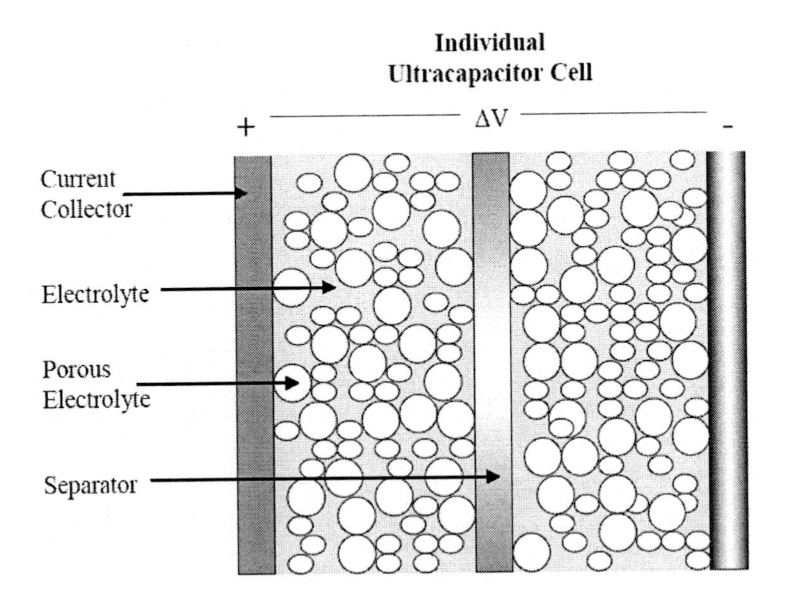

Figure 73: ECDL supercapacitor cell.

Aqueous electrolytes use either acids or bases (H_2SO_4, KOH) but the nominal voltage is limited to 1V. During charging, the electrically charged ions in the electrolyte migrate towards the electrodes of opposite polarity due to the electric field between the charged electrodes created by the applied voltage. Thus two separate charged layers are produced. Although, similar to a battery, the double-layer capacitor depends on electrostatic action. Since no chemical action is involved the effect is easily reversible with minimal degradation in deep discharge or overcharge and the typical cycle life is hundreds of thousands of cycles. Reported cycle life is more than 500,000 cycles at 100% depth of discharge. The limiting factor in terms of lifetime may be the years of operation with reported lifetimes reaching up to 12 years. Another limiting factor is the high self-discharge rate of supercapacitors. This rate is much higher than batteries reaching a level of 14% of nominal energy per month.

Apart from high tolerance to deep discharges, the fact that no chemical reactions are involved means that supercapacitors can be easily charged and discharged in seconds thus being much faster than batteries. Also, no thermal heat or hazardous substances can be released during discharge. Energy efficiency is very high, ranging from 85% up to 98%.

Compared to conventional capacitors, the supercapacitors have significantly larger electrode surface area coupled with much thinner electrical layer between the electrode and the electrolyte. The layer's thickness is only a few molecular diameters. These two attributes mean that supercapacitors have higher capacitances and therefore energy densities than conventional capacitors. Capacitances of 5000F have been reported with supercapacitors and energy densities up to 5Wh/kg compared to 0.5Wh/kg of conventional capacitors. Current currying capability of the supercapacitors is also very high since it is directly proportional to the surface area of the electrodes. Thus, power density of supercapacitors is extremely high, reaching values such as 10000W/kg which is a few orders of magnitude higher than the power densities achieved with batteries. However, due to the low energy density, this high amount of power will only be available for a very short duration. In the cases where supercapacitors are used to provide power for prolonged periods of time, it is at the cost of considerable added weight and bulk of the system due to their low energy density.

Supercapacitor cost is a significant issue for the further commercial use of supercapacitors in industrial applications. Cost, which is estimated to be around 20000$/kWh, is significantly higher compared to well-established storage technologies such as lead-acid batteries. Drastic cost reduction must therefore be accomplished especially in the carbon, electrolyte and separator fields. Currently, the high power storage ability of supercapacitors together with the fast discharge cycles, make them ideal for use in temporary energy storage for capturing and storing the energy from regenerative braking and for providing a booster charge in response to sudden power demands. At the same time, supercapacitors can provide effective short duration peak power boost, and short term peak power back up for UPS applications. By combining a supercapacitor with a battery-based UPS system, the life of the batteries can be extended by load sharing between battery and supercapacitor. The batteries provide power only during the longer interruptions, reducing the peak loads on the battery and permitting the use of smaller batteries.

Currently developments in supercapacitor manufacturing technology have shown that the use of vertically aligned, single-wall carbon nanotubes which are only several atomic diameters in width instead of the porous, amorphous carbon normally employed can significantly increase the supercapacitor capacity and power density. This is due to the fact that the surface area of the electrodes is dramatically increased by the use of such materials. Energy densities of 60Wh/kg and power densities of 100000W/kg can be achieved with this technology.

Pseudo-capacitors and hybrid capacitors are also promising technologies because they can achieve improved performances where ECDL supercapacitors offered inferior capabilities. Pseudo-capacitors use metal oxides or conducting polymers as electrode material and can achieve higher energy and power densities than ECDL supercapacitors. Metal-oxide supercapacitors use aqueous electrolytes and metal oxides such as ruthenium oxide (RuO_2), iridium oxide and nickel oxide mainly for military applications. These supercapacitors are based on a high kinetics charge transfer at the electrode/electrolyte interface transforming ruthenium oxide into ruthenium hydroxide ($Ru(OH)_2$) leading to pseudo-capacitive behaviour. Metal-oxide supercapacitors are however still very expensive to produce and may suffer from lower efficiencies and lower voltage potential due to the need for aqueous electrolytes. Hybrid supercapacitors can reach even higher energy and power densities than the other supercapacitors without sacrifices in affordability or cyclic stability. They are however still a new and unproven technology and still require more research to better understand and realise their full potential.

7.4. Hydrogen Storage Technologies

Currently there are four main technologies for hydrogen storage out of which two are more mature and developed. These are the hydrogen pressurization and the hydrogen adsorption in metal hydrides. The remaining two technologies that are still in research and technological development phase are the adsorption of hydrogen on carbon nanofibers and the liquefaction of hydrogen.

Pressurized hydrogen technology relies on high materials permeability to hydrogen and to their mechanical stability under pressure. Currently steel tanks can store hydrogen at 200 to 250 bar but present very low ratio of stored hydrogen per unit weight. Storage capability increase with higher pressures but stronger materials are then required. Storage tanks with aluminium liners and composite carbon fibre/polymer containers are being used to store hydrogen at 350bar providing higher ratio of stored hydrogen per unit weight (up to 5%). In order to reach higher storage capability, higher pressures are required in the range of 700bar with the unavoidable auxiliary energy requirements for the compression. Research is currently under way to materials that are adequate for use in such high pressures.

The use of metal hydrides as storage mediums is based on the excellent hydrogen absorption properties of these compounds. These compounds, obtained through the direct reaction of certain metals or metal alloys to hydrogen, are capable of absorbing the hydrogen and restoring it when required. These compounds have a low equilibrium pressure at room temperature (lower than atmospheric pressure) in order to prevent leaks and guarantee containment integrity and a low degree of sensitivity to impurities in the hydrogen stored. The use of such a storage technology is safe, as the pressure remains low, and it is compact because most hydrides have high volume absorption capacities (ratio of the volume of hydrogen stored to the volume of metal used). An example of metal hydride containers is shown in Figure 74.

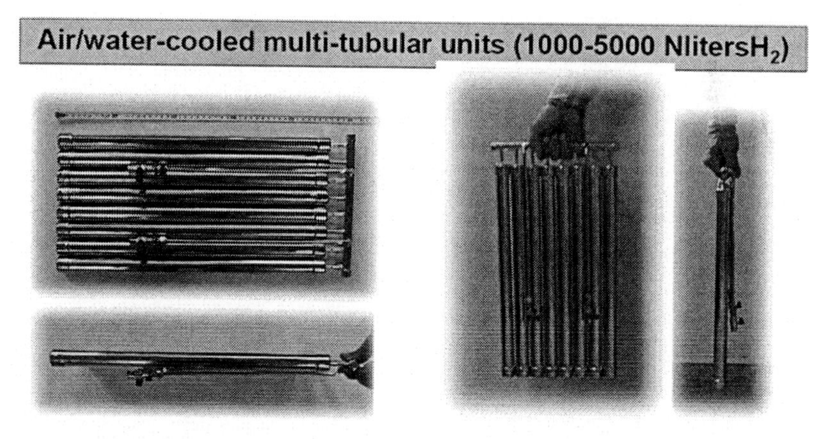

Figure 74. Air/water-cooled multi-tubular metal hydride storage units.

However, metal hydride compounds have some disadvantages. Typically, they exhibit rather low mass absorption capacities (except magnesium hydrides) and do require thermal management system. This is because the absorption of hydrogen is an exothermic reaction (releases heat) while desorption of hydrogen is endothermic. Heating and cooling of the metal hydrides containers is achieved via the use of water running through pipework in the interior of the container. Absorption/desorption kinetics are however very fast in most hydrides thus allowing for fast hydrogen storage and release. Materials most often employed in this hydrogen storage technology are seldom earth materials such as lanthanum, and other materials such as nickel and aluminium.

Liquid hydrogen storage technology use is currently limited. This is due to the properties and cost of the materials used in the manufacturing of the container/tank and the extreme temperatures that are required for such storage. The typical temperature required to maintain hydrogen in a liquid state is around -253°C. Storage containers have to use specific internal liner surrounded by a thermal insulator in order to maintain the required temperature and avoid any evaporation. The whole process is quite inefficient since a lot of energy is used already in the initial stage of hydrogen liquefaction. In addition, liquid hydrogen tanks suffer from leaks (escaping hydrogen flow) due to the unavoidable thermal losses that lead to pressure increase in the tank. This hydrogen self discharge of the tank may reach 3% daily which translates to 100% self discharge in one month.

Finally, the use of carbon nanofibres for hydrogen storage is in its initial research stage. Different materials are still under investigation as to their storage potential depending on the temperature and pressure of the hydrogen.

7.5. Pneumatic Storage Technologies

Pneumatic storage technologies can use either compressed air or compressed gas to achieve energy storage. In compressed gas applications, a system similar to a hydraulic accumulator is employed which can store and release energy through its integration with a motor/generator and a pump/motor. A hydraulic accumulator is a pressure storage device made up of a reservoir in which a non-compressible hydraulic fluid is held under pressure by compressed gas. This technology is typically referred to as "Battery with Oil-Hydraulics and Pneumatics" or BOP. Compressed air applications are able to store energy on a much bigger scale and are used under the abbreviation CAES (Compressed Air Energy Storage). In this type of storage technology, air is stored in pre-specified underground locations and released under pressure to a drive the gas turbines of the electricity generation plant. Both these technologies are explained briefly below.

7.5.1. Liquid-Piston Technology

This technology is based on a compressed gas hydraulic accumulator together with the addition of a pump/motor and a motor/generator as shown in Figure 76. This technology provides an almost perfect isothermal behavior due to the low speed of the compression/expansion process which is distributed over all the storage vessels and hence linked to an enormous heat exchange surface.

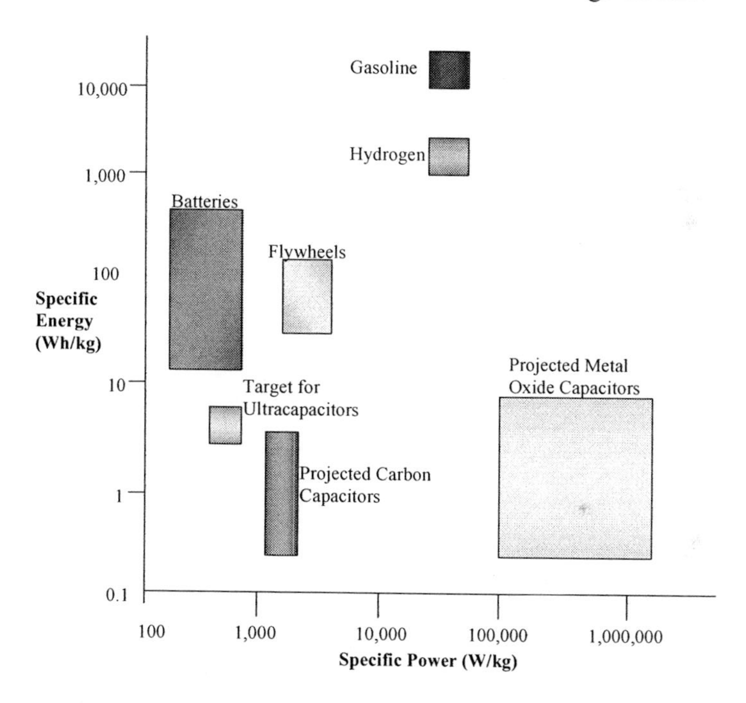

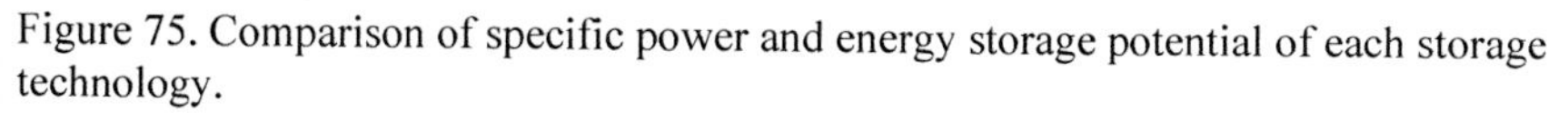
Figure 75. Comparison of specific power and energy storage potential of each storage technology.

The basic principle of operation is that compression and expansion of a trapped volume of gas -usually nitrogen- (see Figure 76) take place in a storage vessel and its volume and pressure are modulated by the amount of fluid/liquid in the vessel. Gas pressures vary from 100bar (no fluid present) to 250 bar (50% of the vessel filled with fluid). During energy storage, the high-efficiency fixed displacement pump/motor is energized by the electrical machine or motor/generator and acts as a pump compressing the gas in the accumulator vessels with the fluid. During discharge of energy, the compressed gas is expanded and fluid is expelled from the vessels to the pump/motor which acts now as a motor to drive the electrical machine as a generator of electric power. To complete the system, a solenoid powered 4-way spool valve is used which works in conjunction with a flywheel, designed to maintain a low-rippled speed for the motor/generator. A fluid expansion reservoir is also necessary.

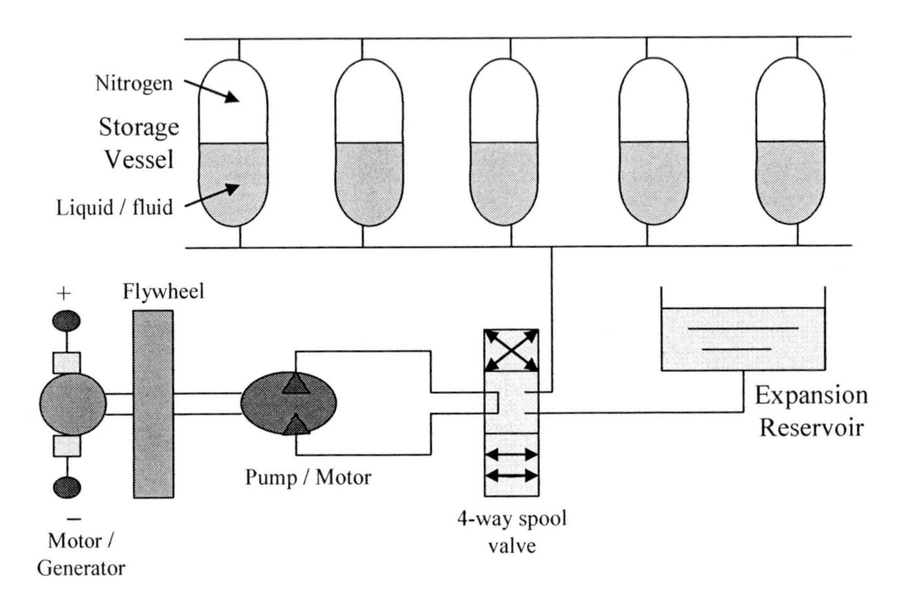

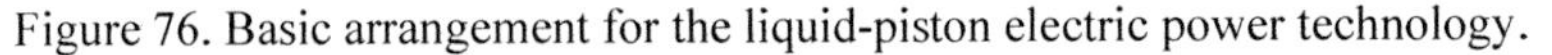
Figure 76. Basic arrangement for the liquid-piston electric power technology.

This technology is not yet on a commercial basis, but it is hoped that it can initially substitute the lead-acid batteries in certain stand-alone stationary equipment applications. The first targeted applications are systems with drives mainly for workshops, food processing, milking and UPS installations. In these types of applications, its performance can compete with the performance of lead-acid batteries although this technology remains more expensive.

Pneumatic storage technology's main advantages over the lead-acid batteries are (a) unlimited cycling ability and lifetime since cycling is not dependent on charge/discharge profiles, (b) lower maintenance required, (c) storage capacity not affected by age or speed of charge, (d) undamaged performance in cases of full discharge, (e) overcharge is impossible, as the system is protected by a relief valve in the hydraulic circuit, (f) free of charge regulators and low voltage cut-off switches, (g) power supply is not linked to capacity and it is limited only by the transformer design characteristics and (h) self discharge at open circuit (tightly closed circuit valve, shut-off system) is almost zero allowing for many years of operation

Some disadvantages of the above technology compared to the lead-acid batteries are (a) energy density is lower and ranges from 3.2Wh/kg to 5.55Wh/kg (250 bar pressure), in fact energy density increases with pressure, however there is limitation to the maximum pressure that is bearable by the hydraulic system components and materials, (b) although self-discharge is almost zero at tightly closed valve, there is considerable self-discharge at voltage standby mode which presents an operational limitation (applications with long shut-off periods, such as seasonal systems, are therefore a more adequate target for pneumatic storages), (c) the chance of possible leakage in the pneumatic and hydraulic piping that needs to be monitored and fixed and (d) energy efficiency is slightly lower to a new lead-acid battery, at around 73% (efficiency level is projected to increase by the advancements in research and by the continuous technological developments in this technology).

7.5.2. Compressed Air Energy Storage

In compressed air energy storage (CAES), off-peak power is taken from the grid and is used to pump air into a sealed underground cavern to a high pressure. The pressurized air is then kept underground for peak use. When needed, this high pressure can drive turbines as the air in the cavern is slowly heated and released. The resulting power may be used at peak hours.

There are many geologic formations that can be used in this used in this scheme. These include naturally occurring aquifers, solution-mined salt caverns and constructed rock caverns. In general, rock caverns are about 60% more expensive to mine than salt caverns for CAES purposes. This is because underground rock caverns are created by excavating solid rock formations, whereas salt caverns are created by solution mining of salt formations.

Aquifer storage is by far the least expensive method and is therefore used in most of the current locations. The other approach to compressed air storage is called CAS, compressed air storage in vessels. In a CAS system, air is stored in fabricated high-pressure tanks. However, the current technology is not advanced enough to manufacture these high-pressure tanks at a feasible cost. The scales proposed are also relatively small compared to CAES systems.

There are five aboveground components required by a basic CAES installation as shown in Figure 77:

- The motor/generator which employs clutches to provide for alternate engagement to the compressor or turbine trains

- The air compressor which may require two or more stages, intercoolers and aftercoolers to achieve economy of compression, and reduce the moisture content
- The recuperator, turbine train, high and low pressure turbines
- Equipment control centre for operating the combustion turbine, compressor, and auxiliaries and to regulate and control changeover from generation mode to storage mode
- Auxiliary equipment consisting of fuel storage and handling, and mechanical and electrical systems to support various heat exchangers required

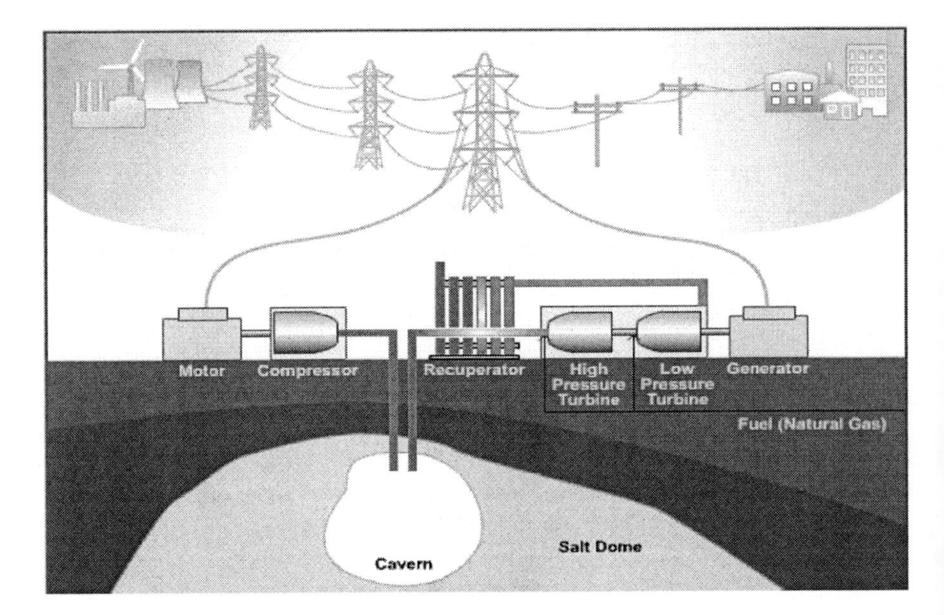

Figure 77. Basic components for a CAES system.

A CAES operates by means of large electric motor driven compressors that store energy in the form of compressed air in the mine. The compression is done outside periods of peak demand. As part of the compression process, the air is cooled prior to injection to make the best possible use of the storage space available. The air is then pressurized to about 75bar. To return electricity to the consumers, air is extracted from the cavern. It is first preheated in the recuperator. The recuperator reuses the energy extracted by the compressor coolers. The heated air is then mixed with small quantities of oil or gar, which is burnt in the combustor. The hot gas from the combustor is expanded in the turbine to generate electricity.

CAES systems can only be used on very large scales. Unlike other systems considered large-scale, CAES is ready to be used with entire power plants. Apart from the hydro-pump, no other storage method has a storage capacity as high as CAES. Typical capacities for a CAES system are around 50-300MW. The storage period is also the longest due to the fact that the losses are very small. A CAES system can be used to store energy for more than a year.

Fast start-up is also an advantage of CAES. A CAES plant can provide a start-up time of about 9 minutes for an emergency start, and about 12 minutes under normal conditions. By comparison, conventional combustion turbine peaking plants typically require 20 to 30 minutes for a normal start-up.

If a natural geological formation is used (rather than CAS), CAES has the advantage that it doesn't involve huge, costly installations. Moreover, the emission of greenhouse gases is substantially lower than in normal gas plants.

The main drawback of CAES is probably the geological structure reliance. There is actually not a lot of underground cavern around, which substantially limits the usability of this storage method. However, for locations where it is suitable, it can provide a viable option for storing energy in large quantities and for long times.

7.6. Pumped Storage Technology

Pumped storage hydroelectricity is a method of storing and producing electricity to supply high peak demands by moving water between reservoirs at different elevations. At times of low electricity demand, excess generation capacity is used to pump water into the higher reservoir (Figure 78). When there is higher demand, water is released back into the lower reservoir through a turbine, generating electricity. Reversible turbine/generator assemblies act as pump and turbine. Some facilities use abandoned mines as the lower reservoir, but many use the height difference between two natural bodies of water or artificial reservoirs. Pure pumped-storage plants just shift the water between reservoirs, but combined pump-storage plants also generate their own electricity like conventional hydroelectric plants through natural steam-flow. Plants that do not use pumped-storage are referred to as conventional hydroelectric plants.

Taking into account evaporation losses from the exposed water surface and conversion losses, approximately 70% to 85% of the electrical energy used to pump the water into the elevated reservoir can be regained. The technique is currently the most cost-effective means of storing large amounts of electrical energy on an operating basis, but capital costs and the presence of appropriate geography are critical decision factors. The relatively low energy density of pumped storage systems requires either a very large body of water or a large variation in height. The only way to store a significant amount of energy is by having a large body of water located on a hill relatively near, but as high as possible above a second body of water. In some places this occurs naturally, in others one or both bodies of water have been man-made.

Pumped storage system may be economical because it flattens out load variations on the power grid, permitting thermal power stations such as coal-fired plants and nuclear plants that provide base-load electricity to continue operating at peak efficiency, while reducing the need for peaking power plants that use costly fuels. Capital costs for purpose-build hydro storage are high, however.

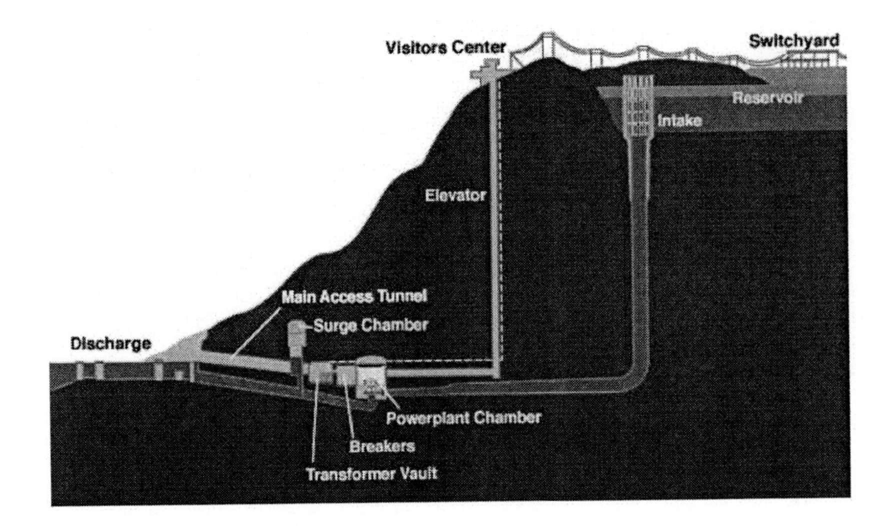

Figure 78. A typical pumped-storage plant.

Along with energy management, pumped storage systems help control electrical network frequency and provide reserve generation. Thermal plants are much less able to respond to sudden changes in electrical demand, potentially causing frequency and voltage instability. Pumped storage plants, like other hydroelectric plants, can respond to load changes within seconds.

7.7. OVERALL COMPARISON

A comparison of the main attributes of three important power storage technologies is presented in Table 17.

Table 17. Comparison among three prominent energy storage systems

Parameter	Lead-acid battery	Flywheel technology	Nickel-Cadmium battery
Storage mechanism	Chemical	Mechanical	Chemical
Life (years in service)	3-12	>20	15-20
Life (deep cycles)	<1500 cycles	$<10^7$ cycles	<3000 cycles
Self-discharge rate	Very Low	Very High	Very Low
Technology	Proven	Promising (Proven)	Proven
Tolerance of overcharge and deep-discharge	Very Low	High	Low
Environmental concerns	Chemical Disposal issues	Slight	Chemical Disposal issues
Energy density	30Wh/kg	5Wh/kg steel 100Wh/kg composite	15-50Wh/kg
Power density	180W/kg	1000W/kg composite	50-1000W/kg
Price / kWh	\$50-\$100	\$400-\$800	\$400-\$2400

Flywheel storage technology is compared to the lead-acid and the nickel-cadmium battery technologies. It is clear from this table that conventional flywheel technology (with steel rotor) differs substantially from battery technology. Although conventional flywheel technology can provide much higher deep-cycle life and therefore more charge-discharge cycles in a storage application, its much higher self-discharge rate is still prohibitive for large scale penetration of flywheel technology. In contrast to a battery, conventional flywheel technology can be used for high power density storage applications, while batteries are only suited to high energy storage applications. However, the newly evolving composite rotor flywheel technology promises higher energy densities comparable to the levels achieved with batteries. Finally in terms of investment cost, flywheel technology incurs higher costs due to the fact that the technology is still at a relatively early stage in its lifecycle compared to traditional battery technologies such as lead-acid or nickel-cadmium batteries.

The deliverable power and energy that can be provided by the majority of the storage technologies discussed so far can be shown in Figure 75. In this Figure, a comparison is presented in terms of energy and power density achievable with each technology. Figure 8 confirms that batteries have the lowest power density, and the highest energy density (with the exception of hydrogen storage technologies). The evolving metal oxide supercapacitors are shown to be able to achieve the highest possible power densities. Compressed air energy storage (CAES) and pumped storage technologies are not shown in Figure 8, since the scale of power applications suitable for these technologies far exceeds the scale of the chart. Typical power applications for these technologies are in the order of 100 of MW.

From the above analysis, it is clear that a large variety of storage technologies exists with each one possessing different attributes and intended for different applications. The choice of the ideal storage technology to be used depends on a number of factors. These are, among others, the amount of energy or power to be stored, the time for which this stored energy is required to be retained or to be released, spacing and environmental constraints, cost, and the exact location of the network on which the storage is required.

It is evident from the above review that batteries are the dominant technology to be used when continuous energy supply is paramount, while technologies such as flywheel and supercapacitors are more suited to power storage applications and where very brief power supply is required such as in uninterrupted power supply requirements. Lithium-ion batteries are becoming increasingly important and have several advantages over the traditional lead-acid batteries. Fuel cells performance is constantly improving in terms of reliability and investment cost, while some types (e.g SOFC) can provide very high efficiencies in the context of combined heat and power (CHP) applications. However, the future penetration of fuel cells remains tied to the high-cost hydrogen production and storage processes. Finally, pumped storage and CAES technologies are suited to very high power, high investment cost generation applications to be used in the transmission system.

REFERENCES

[1] Abu-Khader M.M., Badran O.O., Abdallah S., 2008, "Evaluating multi-axes sun-tracking system at different modes of operation in Jordan", *Renewable and Sustainable Energy Reviews*, 12, 864-873.

[2] Ackemann T., Anderson G., Soder L., 2001, "Distributed generation: a definition", *Electric Power Systems Research*, 57, 195-204.

[3] Anheden M., 2000, "Analysis of gas turbine systems for sustainable energy conversion", *PhD Thesis*, Royal Institute of Technology, Sweden.

[4] Beerbauma S., Weinrebe G., "Solar thermal power generation in India - a techno-economic analysis", *Renewable Energy*, 21, 153-174

[5] Bies D., Prelipceanu A., 2003, "Optimised cooling of the intake air by absorption refrigeration technology – An innovative technology to improve gas turbine performance", *Proceedings of the First International Conference on Industrial Gas Turbine Technologies*.

[6] Bolland O., Stadaas J.F., 1995, "Comparative evaluation of combined cycles and gas turbine systems with water injection, steam injection and recuperation", *Journal of Engineering for Gas Turbines and Power*, 117, 138-145.

[7] Boudries R., Dizene R., 2008, "Potentialities of hydrogen production in Algeria", *International Journal of Hydrogen Energy*, 33, 4476-4487.

[8] Broek M., Faaij A., Turkenburg W., "Planning for an electricity sector with carbon capture and storage: Case of the Netherlands", *International Journal of Greenhouse Gas Control*, 2, 105-129.

[9] Brown S.J., Rowlands I,H, 2009 "Nodal pricing in Ontario, Canada: Implications for solar PV electricity", *Renewable Energy*, 34, 170-178.

[10] Castello P., Tzimas E., Moretto P., Peteves S.D., 2005, "*Techno-economic assesment of hydrogen transmission and distribution systems in Europe in the medium and long term*", European Commission Joint Research Center, Institute for Energy, Report EUR 21586EN.

[11] Cheng D.Y., 1978, "Regenerative parallel compound dual-fluid heat engine", *U.S. Patent 4*,128,994.
[12] Christodoulou C., Karagiorgis G., Poullikkas A., Karagiorgis N., Hadjiargyriou N., 2007, "Green electricity production by a grid-connected H2/fuel cell in Cyprus", *Proceedings of the Renewable Energy Sources and Energy Efficiency.*
[13] Christou C., Hadjipaschalis I., Poullikkas A., 2008, "Assessment of Integrated Gasification Combined Cycle technology competitiveness", *Renewable and Sustainable Energy Reviews*, 12, 2452-2464.
[14] Compaan A.D., 2006, "Photovoltaics: Clean power for the 21st century", Solar Energy Materials and Solar Cells, 90, 2170-2180.
[15] Concentrating solar power – From research to implementation, 2007, European Communities, ISBN 978-92-79-05355-9.
[16] Damen K., Troost M.V., Faaij A., Turkenburg W., 2006, "A comparison of electricity and hydrogen production systems with CO2 capture and storage. Part A: Review and selection of promising conversion and capture technologies", *Progress in Energy and Combustion Science* 32, pp. 215-246.
[17] David J., Herzog H., 2000, "The cost of carbon capture", *Proceedings of the fifth international conference on greenhouse gas control technologies.*
[18] Davison D., 2006, "Performance and costs of power plants with capture and storage of CO2", *Energy 32*, 1163-1176.
[19] Duke R., Williams R., Payne A., 2005, "Accelerating residential PV expansion: demand analysis for competitive electricity markets", *Energy Policy*, 33, 1912-1929.
[20] ec.europa.eu (Energy for a Changing World, European Commission).
[21] El-Khatan W., Salama M.M.A., 2004, "Distributed generation technologies, definitions and benefits", *Electric Power Systems Research*, 71, 119-128.
[22] esolar.cat.com
[23] European Commission, 2006, A vision for zero emissions fossil fuels power plants, *Report by the Zero Emission Fossil Fuel Power Plants Technology Platform*, EUR22043.
[24] European Commission, 2006, A vision for zero emissions fossil fuels power plants, *Report by the Zero Emission Fossil Fuel Power Plants Technology Platform*, EUR22043.
[25] European research on concentrated solar thermal energy, 2004, European Communities, ISBN 92-894-6353-8.
[26] Farina G.L., Bressan L., 1999, "Optimizing IGCC design", *Foster Wheeler Review*, (www.fwc.com/publications).

[27] Feron P.H.M. and Hendriks C.A., 2005, "CO2 Capture Process Principles and Costs", *Oil and Gas Science and Technology – Rev.* IFP, Vol. 60, No. 3, pp. 451-459.

[28] Foster, R.W., 1989, "Turbo STIG – The turbocharged steam injected gas turbine cycle", *ASME paper* 89-GT-100.

[29] Gadde S., Wu J., Gulati A., McQuiggan G., 2006, "Syngas capable combustion systems development for advanced gas turbines", *Proceedings of ASME Turbo Expo 2006,* Paper No: GT2006-90970.

[30] Ganapathy V., Heil B., Rentz J., 1988, "Heat recovery steam generator for Cheng cycle application", *ASME Industrial Power Conference*, PWR, 4.

[31] Garcia G.O., Dougles P., Croiset E., Zheng L., 2006, "Techno-economic evaluation of IGCC power plants for CO2 avoidance", *Energy Conversion and Management,* 47, 2250-2259.

[32] García-Rodrígueza L., Blanco-Gálvezb J., 2007, "Solar-heated Rankine cycles for water and electricity production: POWERSOL project", *Desalination*, 212, 311–318.

[33] Gas Turbine World, 2007, January-February, 37/1, 23-48.

[34] Geosits R.F., Schmoe L.A., 2005, "IGCC – The challenges of integration", *Proceedings of GT2005 ASME Turbo Expo 2005*, Paper No: GT2005-68997.

[35] Green M.A., 2000, "Photovoltaics : technology overview", *Energy Policy,* 28, 989-998.

[36] Greene N., Hammerschlag R., 2000, "Small and Clean Is Beautiful: Exploring the Emissions of Distributed Generation and Pollution Prevention Policies", *The Electricity Journal*, 6, 50-60.

[37] Hadjipaschalis I., Christou C., Poullikkas A., 2008, "Assessment of future sustainable power technologies with carbon capture and storage", *International Journal of Emerging Electric Power Systems*, 9/1, Art. 5.

[38] Hadjipaschalis I., Poullikkas A., Efthimiou V., 2009, "Overview of current and future energy storage technologies for electric power applications", *Renewable and Sustainable Energy Reviews*, 13, 1513-1522.

[39] Hadjipaschalis I., Kourtis G., Poullikkas A., 2009, "Assessment of oxyfuel power generation technologies", *Renewable and Sustainable Energy Reviews,* 13, 2637-2644.

[40] Hellfritsch S., Gampe U., "*Modern coal-fired oxyfuel power plants with CO_2 capture – energetic and economical evaluation*", Dresden University of Technology, Germany.

[41] Heppenstall T., 1998, "Advanced gas turbine cycles for power generation: a critical review", *Applied Thermal Engineering*, 18, 837-846.

[42] Hijikata T., 2002, "Research and development of international clean energy network using hydrogen energy (WE-NET)", *International Journal of Hydrogen Energy*, 27, 115-129.

[43] Hoffmann W., 2006, "PV solar electricity industry: Market growth and perspective", *Solar Energy Materials and Solar Cells*, 90, 3285-3311

[44] Hohlein B., Grube T., Lokurlu A., Stolten D., 2003, "Fuel cells for mobile and stationary applications – Analysis of options and challenges", Proccedings of the World Renewable Energy Congress VII, Fuel Cells Systems, Stolen D., Emonts B. (Eds), 53-66.

[45] Ibrahim H., Ilinca A., Perron J., 2007, "Energy storage systems – Characteristics and comparisons", *Renewable and sustainable Energy Reviews,* 12, 1221-1250.

[46] International Energy Agency (IEA), 2007, "*PVPS Annual Report*", Photovoltaic power systems program (www.iea-pvps.org)

[47] Investigation on Storage Technologies for Intermittent Renewable Energies: Evaluation and recommended R&D strategy, INVESTIRE-NETWORK (ENK5-CT-2000-20336), Storage Technology Reports (www.itpower.co.uk)

[48] IPCC special report on carbon dioxide capture and storage, 2005, Prepared by Working Group III of the Intergovernmental Panel on Climate Change, Cambridge University Press, Cambridge, UK, New York, NY.

[49] IPCC special report on carbon dioxide capture and storage, 2005, Prepared by Working Group III of the Intergovernmental Panel on Climate Change, Cambridge University Press, Cambridge, UK, New York, NY.

[50] Jager-Waldau A., 2002, Status of PV research, solar cell production and market implementation in Japan, USA and the European Union, European Commission, Joint research Centre, EUR 20425 EN.

[51] Janes J., 1996, "A fully enhanced gas turbine for surface ships", *ASME Paper* 96-GT-527.

[52] Jericha H., Gottlich E., Sanz W., Heitmeir F., 2003, "Design optimisation of the Gratz cycle prototype plant", *ASME paper* GT2003-38120.

[53] Jordal K., et al, "*Oxyfuel combustion for coal-fired power generation with CO2 capture - Opportunities and challenges*", Vatenfall Utveckling AB.

[54] Kakaras E., Doukelis A., Scharfe J., 2001, "Applications of gas turbine plants with cooled compressor intake air", *ASME paper* 2001-GT-110.

[55] Kakaras E., et al, 2007, "Economic implications of oxyfuel application in a lignite-fired power plant", *Fuel,* (doi: 10.1016/j.fuel.2007.03.035).

[56] Kakaras E., et al, 2007, "Oxyfuel boiler design in a lignite-fired power plant", *Fuel,* (doi: 10.1016/j.fuel.2007.03.037).

[57] kawasaki.com

[58] Kesser K.F., Hoffman M.A., Baughm J.W., 1994, "Analysis of a basic chemically recuperated gas turbine power plant", *Journal of Engineering for Gas Turbines and Power*, 116, 277-284.

[59] Kolev N., Schaber K., Kolev D., 2001, "A new type of a gas-turbine cycle with increased efficiency, *Applied Thermal Engineering*, 21, 391-405.

[60] Korobitsyn M.A., 1998, "New and advanced energy conversion technologies - Analysis of cogeneration combined and integrated cycles", *PhD Thesis*, University of Twente.

[61] Leo T.J., Perez G.I., Perez N.P., 2003, "Gas turbine turbocharged by a steam turbine: a gas turbine solution increasing combined power plant efficiency and power", *Applied Thermal Engineering*, 23, 1913-1929.

[62] Letendre S.E., Perez R., 2006, "Understanding the Benefits of Dispersed Grid-Connected Photovoltaics: From Avoiding the Next Major Outage to Taming Wholesale Power Markets", *The Electricity Journal*, 19/6, 64-72.

[63] Magnusson M., Smedman A.S., 1999, "Air flow behind wind turbines", *Journal of Wind Engineering and Industrial Aerodynamics*, 80, 169-189.

[64] Marion J., 2004, "Technology options for controlling CO2 emissions from fossil-fuelled power plants", *Proceedings of the Third Annual Conference on Carbon Sequestration*, Alexandria, VA, USA.

[65] Martins F.R., Rüther R., Pereira E.B., Abreu S.L., 2008, "Solar energy scenarios in Brazil. Part two: Photovoltaics applications", *Energy Policy,* 36, 2865-2877

[66] Maurstad O., 2005, "An overview of coal based Integrated Gasification Combined Cycle (IGCC) technology", Massachusetts Institute of Technology, Laboratory for Energy and the Environment, Publication No: LFEE 2005-002 WP (fee.mit.edu/publications).

[67] Mellit A., Kalogirou S.A, Hontoria L., Shaari S., 2008, "Artificial intelligence techniques for sizing photovoltaic systems: A review", *Renewable and Sustainable Energy Reviews*, in press.

[68] Milborrow D., Hartnell G., Cutts N., 2000, *Renewable energy in the EU*, Financial Times Energy, London.

[69] Moliere M., 2000, "Stationary gas turbines and primary energies: A review of fuel influence on energy and combustion performances", *Int. J. Therm. Sci.*, 39, 141-172.

[70] Morisson V., Rady M., Palomo E., Arquis E., 2008, "Thermal energy storage systems for electricity production using solar energy direct steam generation technology", *Chemical Engineering and Processing*, 47, 499–507.

[71] Muntasser M.A., Bara M.F., Quadri H.A., El-Tarabelsi R., La-azebi I.F., 2000, "Photovoltaic marketing in developing countries", *Applied Energy*, 65, 67-72.

[72] Najjar Y.S.H., 2000, "Gas turbine cogeneration systems: a review of some novel cycles", *Applied Thermal Engineering*, 20, 179-197.

[73] Narula R.G., Wen H., Himes K., 2002, "Incremental cost of CO2 reduction in power plants", *Proceedings of ASME Turbo Expo 2002*, Paper No: GT-2002-30259.

[74] Nsakala N.y., et al, 2004, "Oxygen-fired circulating fluidisized bed boilers for greenhouse gas emissions control and other applications", *Proceedings of the Third Annual Conference on Carbon Sequestration*, Alexandria, VA, USA.

[75] Oliver M., Jackson T., 1999, "The market of solar photovoltaics", *Energy Policy*, 27, 371-385.

[76] Pepermans G., Driesen J., Haeseldonckx D., Belmans R., D'haeseleer W., 2004, "Distributed generation : definition, benefits and issues", *Energy Policy*, article in press.

[77] Poullikkas A., 1999, "Compressibility and condensation effects when pumping gas-liquid mixtures", *Fluid Dynamics Research*, 25, 57-62.

[78] Poullikkas A., 2000, "Two phase flow performance of nuclear reactor cooling pumps", *Progress in Nuclear Energy*, 36, 123-130.

[79] Poullikkas A., 2001, "A technology selection algorithm for independent power producers", *The Electricity Journal*, 14/6, 80-84.

[80] Poullikkas A., 2001, "Optimization algorithm for reverse osmosis desalination economics", *Desalination*, 133, 75-81.

[81] Poullikkas A., 2003, "Effects of two-phase liquid-gas flow on the performance of nuclear reactor cooling pumps", *Progress in Nuclear Energy*, 42, 3-10.

[82] Poullikkas A., 2004, "Cost-benefit analysis for the use of additives in heavy fuel oil fired boilers", *Energy Conversion and Management*, 45, 1725-1734.

[83] Poullikkas A., 2004, "Parametric study for the penetration of combined cycle technologies into Cyprus power system", *Applied Thermal Engineering*, 24, 1675-1685.

[84] Poullikkas A., 2005, "An overview of current and future sustainable gas turbine technologies", *Renewable and Sustainable Energy Reviews*, 9, 409-443.

[85] Poullikkas A., 2005, "Operating cost and water economy of mixed air steam turbines", *Applied Thermal Engineering*, 25, 1949-960.

[86] Poullikkas A., 2005, "Technical and economic analysis for the integration of small reverse osmosis desalination plants into MAST gas turbine cycles for power generation", *Desalination,* 172, 145-150.
[87] Poullikkas A., 2006, "Implementation of MAST gas turbine technologies for large scale power generation", *Energy Sources*, Part A, 28, 1433-1446.
[88] Poullikkas A., 2006, I.P.P. ALGORITHM v2.1, Software for power technology selection in competitive electricity markets, © 2000 – 2006, User Manual.
[89] Poullikkas A., 2007, "Implementation of distributed generation technologies in isolated power systems", *Renewable and Sustainable Energy Reviews*, 11, 30-56.
[90] Poullikkas A., 2008, "Optimization procedures for the selection of reverse osmosis desalination plants", Chapter 16 in *Desalination Research Progress*, NOVA Science Publishers, Inc., New York, ISBN: 978-1-60456-567-6, 449-478.
[91] Poullikkas A., 2008, "Performance of nuclear reactor cooling pumps under two phase liquid-gas flow conditions", Chapter 6 in *Nuclear Energy Research Progress*, NOVA Science Publishers, Inc., New York, ISBN: 978-1-60456-365-8, 175-194.
[92] Poullikkas A., 2009, "A decouple optimization method for power technology selection in competitive markets", *Energy Sources*, Part B, 4, 199-211.
[93] Poullikkas A., 2009, "Economic analysis of power generation from parabolic trough solar thermal plants for the Mediterranean region – A case study for the island of Cyprus", *Renewable and Sustainable Energy Reviews*, 13, 2474-2484.
[94] Poullikkas A., 2009, "Parametric cost-benefit analysis for the installation of photovoltaic parks in the island of Cyprus", *Energy Policy*, 37, 3673-3680.
[95] Poullikkas A., Hadjipaschalis I., Christou C., 2009, "The cost of integration of zero emission power plants – A case study for the island of Cyprus", *Energy Policy*, 37, 669-679.
[96] Poullikkas A., Kellas A., 2004, "The use of sustainable combined cycle technologies in Cyprus: A case study for the use of LOTHECO cycle", *Renewable and Sustainable Energy Reviews*, 8, 521-544.
[97] Rao A.D., 1989, "*Process for producing power*", US Patent 4,829,763.
[98] Rao A.D., Yi Y., Samuelsen G.S., 2003, "Gas Turbine Based High Efficiency "Vision 21" Natural Gas and Coal Central Plants", *Proceedings of the First International Conference on Industrial Gas Turbine Technologies.*

[99] Rezvani S., et al, 2007, "Comparative assessment of sub-critical versus advanced super-critical oxyfuel fired PF boilers with CO2 sequestration facilities", *Fuel,* (doi:10.1016/j.fuel.2007.01.027).

[100] Rice I.G., 1995, "Steam injected gas turbine analysis: Steam rates", *Journal of Engineering for Gas Turbines and Power,* 117, 347-353.

[101] Rosenberg G.W., Dwight C.A., Walker M.R., 2004, "*Financing IGCC – 3 party covenant*", Harvard University, School of Government, (www.bcsia.harvard. edu.energy).

[102] Rubin E. S., Chen C., Rao A., 2007, "Cost and performance of fossil fuel power plants with CO2 capture and storage", *Energy Policy*, article in press (doi:10.1016/jenpol.2007.03.009).

[103] Rubin E. S., Chen C., Rao A., 2007, "Cost and performance of fossil fuel power plants with CO2 capture and storage", *Energy Policy*, article in press (doi:10.1016/jenpol.2007.03.009).

[104] Rubin E. S., Rao A., Chen C., 2004, "Comparative Assessments of fossil fuel power plants with CO2 capture and storage", *Proceedings of 7th International Conference on Greenhouse gas control Technologies* (GHGT-7), Vancouver, Canada.

[105] Rubin E., et al, 2007, "Use of experience curves to estimate the future cost of power plants with CO2 capture", *International Journal of Greenhouse Gas Control* I, 188-197.

[106] Salas V., Olias E., 2008, "Overview of the photovoltaic technology status and perspective in Spain" *Renewable and Sustainable Energy Reviews*, in press

[107] Šály V., Ružinský M., Baratka S., 2006, "Photovoltaics in Slovakia—status and conditions for development within integrating Europe", *Renewable Energy*, 31, 865-875

[108] Sark W.G.J.H.M., Brandsen G.W., Fleuster M., Hekkert M.P., 2007, "Analysis of the silicon market: Will thin films profit?", *Energy Policy*, 35, 3121-3125

[109] Sekkappan G., Panesar R., Hume S., 2007, "Oxyfuel CO2 capture for pulverized coal – an evolutionary approach" Third International Conference on Clean Coal Technologies for our Future, Cagliari, Sardinia.

[110] Shum K.L., Watanabe C., 2008, "Towards a local learning (innovation) model of solar photovoltaic deployment", *Energy Policy*, 36, 508-521

[111] Sinclair M., 2008, "Mainstreaming Solar PV in the USA", *Renewable Energy Focus*, 9, 64-70.

[112] Sinha P., Kriegner C.J., Schew W.A., Kaczmar S.W., Traister M., Wilson D.J., 2008, "Regulatory policy governing cadmium-telluride photovoltaics:

A case study contrasting life cycle management with the precautionary principle", *Energy Policy*, 36, 381-387.
[113] Spanggaard H., Krebs F.C., 2004, "A brief history of the development of organic and polymeric photovoltaics", *Solar Energy Materials and Solar Cells,* 83, 125-146.
[114] Spero C., et al, 2008, "*Callide Oxyfuel Project*", IEAGHG International Oxy-Combustion Network, Yokohama, Japan.
[115] Stambouli A.B., Traversa E., 2002, "Fuel cells, an alternative to standard energy sources", *Renewable and Sustainable Energy Reviews*, 6, 297-306.
[116] Supercapacitors and other Nano-enabled energy systems, Marin Halper, MITRE Nanosystems Group, 2006 (www.mitre.org).
[117] Tazaki M., Sugimoto T., Sambonsugi K., 2003, "*Steam injection type gas turbine*", US Patent 6,502,403.
[118] Toporov D., Forster M., Kneer R., 2007, "Combustion of pulverized fuel under oxycoal conditions at low oxygen concentrations", Third International Conference on Clean Coal Technologies for our Future, Cagliari, Sardinia.
[119] Traverso A., Massardo A. F., 2002, "Thermodynamic analysis of mixed gas-steam cycles", *Applied Thermal Engineering*, 22, 1-21.
[120] Wang F.J., Chiou J.S., 2004, "Integration of steam injection and inlet air cooling for a gas turbine generation system", *Energy Conversion and Management,* 45, 15-26.
[121] Williams M.C., Strakey J.P., Singhal S.C., 2004, "U.S. distributed generation fuel cell program", *Journal of Power Sources*, 131, 79-85.
[122] www.activepower.com
[123] www.cachetco2.com
[124] www.cap-xx.com
[125] www.cie.org.cy
[126] www.desertec.org
[127] www.directindustry.com
[128] www.distres.eu
[129] www.doc.ic.ac.uk
[130] www.eac.com.cy
[131] www.electricitystorage.org
[132] www.enercon.com
[133] www.energy.ca.gov
[134] www.eupvplatform.org
[135] www.fe.doe.gov
[136] www.gas-turbines.com
[137] www.gepower.com

[138] www.heron.nl/index.htm
[139] www.hystoretechnologies.com
[140] www.mpoweruk.com
[141] www.nrel.gov
[142] www.oit.doe.gov
[143] www.power.alstom.com
[144] www.renewables-made-in-germany.com/en/solar-thermal-power-plants
[145] www.roll-royce.com/index-flash.jsp
[146] www.siemenswestinghouse.com
[147] www.vok.lth.se/index-uk.html
[148] www.worldbank.org
[149] www.zero-emissionplatform.eu
[150] www.zero-emissionplatform.eu
[151] www2.rolls-royce.com
[152] Zondag H.A., 2008, "Flat-plate PV-Thermal collectors and systems: A review", *Renewable and Sustainable Energy Reviews*, 12, 891-959.
[153] Zwaan. B, Rabl A., 2003, "Prospects for PV: a learning curve analysis", *Solar Energy*, 74, 19-31.

INDEX

A

B

C

D

E

F

G

H

I

N

O

P

Q

R

S